ENGINEERING APPLICATIONS
OF
LASERS AND HOLOGRAPHY

OPTICAL PHYSICS AND ENGINEERING

Series Editor: **William L. Wolfe**
Optical Sciences Center, University of Arizona, Tucson, Arizona

M. A. Bramson
Infrared Radiation: A Handbook for Applications

Sol Nudelman and S. S. Mitra, Editors
Optical Properties of Solids

S. S. Mitra and Sol Nudelman, Editors
Far-Infrared Properties of Solids

Lucien M. Biberman and Sol Nudelman, Editors
Photoelectronic Imaging Devices
    Volume 1: Physical Processes and Methods of Analysis
    Volume 2: Devices and Their Evaluation

A. M. Ratner
Spectral, Spatial, and Temporal Properties of Lasers

Lucien M. Biberman, Editor
Perception of Displayed Information

W. B. Allan
Fibre Optics: Theory and Practice

Albert Rose
Vision: Human and Electronic

J. M. Lloyd
Thermal Imaging Systems

Winston E. Kock
Engineering Applications of Lasers and Holography

Shashanka S. Mitra and Bernard Bendow, Editors
Optical Properties of Highly Transparent Solids

A Continuation Order Plan is available for this series. A continuation order will bring delivery of each new volume immediately upon publication. Volumes are billed only upon actual shipment. For further information please contact the publisher.

ENGINEERING APPLICATIONS OF LASERS AND HOLOGRAPHY

Winston E. Kock

Acting Director
The Herman Schneider Laboratory of
Basic and Applied Science Research
University of Cincinnati
Cincinnati, Ohio

PLENUM PRESS · NEW YORK AND LONDON

Library of Congress Cataloging in Publication Data

Kock, Winston E
  Engineering applications of lasers and holography.

  (Optical physics and engineering)
  Edition for 1969 published under title: Lasers and holography.
  Bibliography: p.
  Includes index.
  1. Holography. 2. Lasers. I. Title.
  QC449.K6 1975   621.36'6   75-17507
  ISBN 0-306-30849-5

© 1975 Plenum Press, New York
A Division of Plenum Publishing Corporation
227 West 17th Street, New York, N.Y. 10011

United Kingdom edition published by Plenum Press, London
A Division of Plenum Publishing Company, Ltd.
Davis House (4th Floor), 8 Scrubs Lane, Harlesden, London, NW10 6SE, England

Printed in the United States of America

Nobel Laureate Dennis Gabor, who
originally conceived of the idea of holography

PREFACE

This book is intended for upperclass college students as an introduction to the growing field of coherent optics and to the increasing number of its applications, and also for those versed in other fields who wish to gain perspective and insight without detailed calculations. It is an outgrowth of the author's Science Study Series book *Lasers and Holography*.* Besides being an updated and expanded version of that book, it includes discussions of numerous recent applications. It differs in its slightly higher analytical level and in the inclusion of large numbers of references, which enable the reader to obtain further information on subjects of interest to him. The level was selected to match the capabilities of students in their middle college years so as to permit them to make an early assessment of possible career interests in any of the many interdisciplinary fields now embracing the technologies of modern optics.

It is hoped that the book can be used (as has occurred rather extensively with another of the author's Science Study Series books, *Sound Waves and Light Waves†*) as an auxiliary reading assignment for students in various disciplines. The author strongly believes that the promise of continued growth in this field, as evidenced by the extensive participation in technology developments by industry, both within the U.S. and abroad, identifies the subject as

---

* Doubleday, 1969 (hard cover and paperback). This book was chosen by Heinemann Education Books, Ltd. for No. 39 in their Science Series (1972, hard cover), by Editorial Universitaria De Buenos Aires (EUDEBA) for No. 42 in their New Science Collection (1972, paperback), by World Publishing (MIR, Moscow) for their Science and Technology Series (1971, paperback), and by Kawade Shobo (Tokyo) for No. 41 in their Science Series (1971, paperback).

† Doubleday, 1965, also widely translated: No. 109 in the Springer Verstandliche Wissenschaft series (1971, paperback), No. 17 in the Zanichelli Scientific Monograph series (Rome, 1966, paperback), No. 27 in the Kawade Shobo Science Series, (Tokyo, 1969, paperback), and in the World Publishing Science and Technology Series (MIR, Moscow, 1966, paperback).

one of more than average significance to engineering students, and the auxiliary reading assignment procedure would enable the student to become aware of the future importance of this new technology. Several chapters include an introductory review of the history and growth of the field discussed so as to further permit the reader to recognize the impact that modern optics is making in that area.

ACKNOWLEDGMENTS

The author is indebted to F. K. Harvey of the Bell Telephone Laboratories for the photographs portraying microwaves and sound waves, to Lowell Rosen and John Rendiero of the NASA Electronics Research Center for various hologram photographs, and to the Bendix Corporation, the Bell Telephone Laboratories, and numerous corporations for various photos of equipment and records. Numerous sections of the book are based on the author's presentations at the three U.S.–Japan Seminars on Holography, held in Tokyo, Washington, D.C., and Honolulu, Hawaii in 1967, 1969, and 1973, respectively, under the auspices of the National Science Foundation and the Japanese Society for the Promotion of Science. The author wishes to thank Professor George Stroke, State University of New York, Head of the U.S. delegations, for the invitations to participate as one of the U.S. delegates and as co-chairman for the last two seminars.

CONTENTS

ix

Chapter 6
LASER FUNDAMENTALS 103

Chapter 7
RECENT DEVELOPMENTS IN LASERS 123

INTRODUCTION

Holography and photography are two ways of recording, on film, information about a scene we view with our eyes. The basic mechanisms by which they accomplish their purpose, and the images which result, however, are quite different. As the words *holo* (complete) and *gram* (message) connote, the hologram captures the entire message of the scene in all its visual properties, including the realism of three dimensions. The photograph, on the other hand, collapses into one plane, the plane of the print, all the scenic depth we perceive in the actual scene.

The Hungarian-born British scientist Dennis Gabor conceived of holography in 1947 as a new method for photographically recording a three-dimensional image. As is true of many exceptional ideas, it is hard to understand why holography had not been thought of sooner. As we shall see, it involves merely the process of photographically recording the pattern formed by two interfering sets of light waves, one of these wave sets being a *reference wave*. When Gabor conceived this idea, the special kind of light (a single-frequency form called coherent light) needed to demonstrate the full capabilities of holography was not readily available. It became available only after the *laser*, a new light source first demonstrated in 1960, was developed. An atomic process called *stimulated emission* is responsible for the light generated in a laser, and the name laser is an acronym formed from the words *l*ight *a*mplification by the *s*timulated *e*mission of *r*adiation. Because the light sources available to Gabor in 1947 could not demonstrate it fully, holography lay almost dormant for many years. In 1963, the American scientist Emmett N. Leith introduced the laser to holography. The subsequent advances made by him, by another American scientist, George W. Stroke, and by their many co-workers, led to a tremendous explosion in holography development.

A hologram records the interference pattern formed by the combination

1

of a reference wave with the light waves issuing from a scene, and when this photographic record is developed and again illuminated with laser light, the original scene is presented to the viewer as a reconstructed image. This image manifests such vivid realism that the viewer is tempted to reach out and try to touch the objects of the scene. The hologram plate itself resembles a window with the imaged scene appearing behind it in full depth. The viewer has available to him many views of the scene; to see around an object in the foreground, he simply raises his head or moves it left or right. This is in contrast to the older, two-photograph stereopictures, which provide an excellent three-dimensional view of the scene, but only one view. Also, photography uses lenses, and lenses allow only objects at a certain distance from the camera to be in truly sharp focus. In the hologram process, no lenses are used, and all objects near and far, are portrayed in its image in extremely sharp focus.

Gabor was the first to use the term hologram, and Stroke proposed the term holography. Both have become the generally accepted terms, with the first now generally considered as the record, the second as the process. To give the reader a general overview of holography, the first chapters review the underlying wave concepts including coherence, diffraction, and interference. Next, the nature of the holograms themselves is discussed, followed by a description of recent developments in lasers. The last eight chapters describe some of the many fields in which the two techniques are finding applications. A list of suggested further readings is given at the back of the book.

1

FUNDAMENTAL WAVE PROPERTIES

Certain properties of wave motion are manifested by water waves, such as those formed when a pebble is dropped on the surface of a still pond, as shown in Figure 1.1. Because all wave energy travels with a certain speed, such water waves move outward with a wave speed or *velocity of propagation*,

FIGURE 1.1. Water waves on a pond. The velocity of propagation is v, the distance from crest to crest is the wavelength λ, and the periodicity of the up-and-down motion of a point on the surface is the frequency f (or ν).

v. Waves also have a *wavelength*, the distance from crest to crest; it is usually designated, as shown, by the Greek letter λ. If, in Figure 1.1, we were to position one finger so that it just touched the crests of the waves, we could feel each crest as it passes by. If the successive crests are widely separated, they touch our finger less often, less frequently, than if the crests are close together. The expression *frequency* is therefore used to designate how frequently (how many times in one second) the crests pass a given point. The velocity, the wavelength, and the frequency [stated as cycles per second or hertz (Hz)] are thus related by

$$f = v/\lambda \tag{1.1}$$

This says simply that the shorter the wavelength the more frequently the wave crests pass a given point, and similarly, the higher the velocity, the more frequently the crests pass. No proportionality constant is needed in Equation (1.1) if the same unit of length is used for both the wavelength and the velocity and if the same unit of time (usually the second) is used for both the frequency and the velocity.

1.1 Wave Interference

If a simple and uniform set of waves, as in Figure 1.1, were to meet a second set of similarly uniform, single-wavelength waves, a phenomenon called *interference* would result. (The wave fields add algebraically.) At certain points, they reinforce (a condition called *constructive* interference), and at others they cancel (a condition called *destructive* interference). As shown at the left of Figure 1.2, when the *crests* of one wave set, *A*, coincide with the *crests* of a second set, *B*, constructive interference occurs, and the height of the combined crests increases. When, on the other hand, the *crests* of one source coincide with the *troughs* of the second source, as shown on the right,

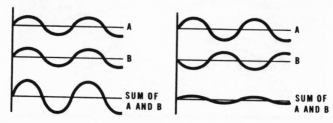

FIGURE 1.2. Two waves of the same wavelength add (at the left) if their crests and troughs coincide, and subtract (at the right) if the crest of one coincides with the trough of the other.

destructive interference occurs, and the combined crest height is lowered. For sound waves, such additive and subtractive effects cause increases and decreases in loudness in the sound pattern; for light waves, they cause variations in brightness or light intensity.

1.2. Phase

By equating one full wavelength to the 360° of a circle, we arrive at the concept of phase. Thus, in Figure 1.2, a negative or positive *phase shift* of 180° would correspond to one wave being shifted to the right or left by one half wavelength. Since the horizontal ordinate of Figure 1.2 represents the passing of time, a shift along that direction is a shift or delay in time.

When the waves are not single frequency, the concept of phase is not applicable, and then only the time-delay concept is useful.

1.3. Wave Visualization

Many wave phenomena can be easily understood if the space patterns of the waves are delineated. To make *sound* patterns visible, a microphone is used (Figure 1.3); it senses the varying sound pressure of the sound wave and converts this varying pressure into a varying electrical current. This electrical current is made visible by amplifying it and causing it to light a neon lamp.[1] Then, in the presence of a loud sound the neon lamp will be brightly lit, whereas for a weaker sound the neon lamp will be dim. The method of photographically recording this brightness pattern is shown in Figure 1.3. The camera at the left is set at time exposure and aimed at the area of interest. The microphone–light combination is made to scan the area, with the camera recording the light intensity variations from spot to spot. In the center is the scanning rod; it has a small microphone and a neon bulb affixed to its end. The white object at the upper right is an acoustic lens; the dark hole behind it is the mouth of a horn, from which the sound waves originate. An electric motor produces the up-and-down motion of the microphone, and it also causes the entire assembly to move slowly toward the reader.

Figure 1.4 shows the result of a time exposure of the light from this device after it has traversed a steady sound field existing in front of a long horn or megaphone. Megaphones cause a person's voice to be directed more strongly in a desired direction; the figure shows this effect. The brighter areas are

FIGURE 1.3. A pickup and light combination can scan an area of interest, permitting a space pattern of sound or microwaves to be recorded photographically by the camera set at time exposure.

indicative of the fact that much stronger sound intensities exist in the direction in which the horn is pointed.

The fact that addition and cancellation effects occur at half-wavelength intervals permits us to exploit these interference effects for portraying wave *motion.*[2] The procedure for doing this is shown in Figure 1.5. Near the bottom of the figure is an electrical connection which causes some direct signal from the oscillator to be sent to the microphone amplifier. There it is combined with the microphone signal received by it from the telephone handset. These two signals can be envisioned as being the waves *A* and *B* of Figure 1.2. Whereas the direct, i.e., wire-line, signal is a constant one (corresponding to wave *A* in Figure 1.2), the scanning microphone (and neon lamp) changes its distance from the telephone sound source. Constructive and destructive interference occurs, with each repeating itself at one-wavelength intervals (i.e., whenever wave *B* of Figure 1.2 has become shifted by one or more complete wavelengths). At points in space that are one, two, or any integral number of

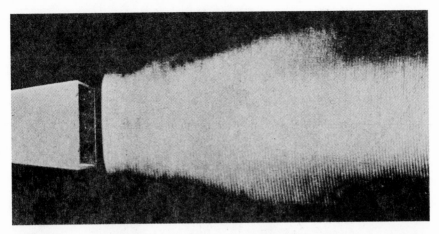

FIGURE 1.4. The pattern of sound intensity generated by the horn loudspeaker at the left is portrayed here (as recorded by the apparatus shown in Figure 1.3). The horn aperture is 6 in. and the frequency is 9 kHz.

wavelengths distant from the telephone sound source, the two signals add, and the neon lamp is bright. At half-wavelength points, the lamp is dim. This effect is shown in Figure 1.6. It shows that single-tone sound waves, like water waves, generate a moving pattern of circular waves centered at the telephone receiver. In the figure, the telephone receiver is reproducing a pure tone signal of 4000 Hz. Because the receiver dimension is small in terms of

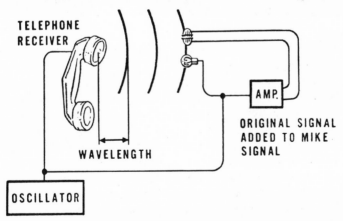

FIGURE 1.5. Combining the radiated signal from the telephone receiver (as picked up by the scanning microphone–lamp combination) with a *direct* signal from the oscillator source generates wave addition and subtraction as the combination moves away from the receiver sound source.

FIGURE 1.6. The process of wave addition and subtraction (Figure 1.5) permits the actual wave fronts of the radiated sound waves to be portrayed. At a frequency of 4000 Hz a telephone receiver is nondirectional and the sound waves spread out in all directions.

the sound wavelength involved, it is not as directional as the megaphone horn of Figure 1.4, and the waves, therefore, spread out in a fairly uniform circular pattern.

1.4. Diffraction

Diffraction effects occur whenever waves strike opaque objects, and wave theory predicts that for certain situations some rather unusual phenomena take place. One such phenomenon, described by Lord Rayleigh, involves the presence of a bright spot in the *shadow* of an opaque disk upon which light waves are falling. From our everyday observations of the shadows cast by opaque objects, we would find it difficult to believe that the wave intensity or brightness at a point in the deep shadow of a disk is, as Lord Rayleigh stated in his *Theory of Sound*,[3] "the same as if no obstacle at all were interposed."

We will first examine what takes place in the shadow of a horizontal board (a "knife edge"). This pattern is shown in Figure 1.7. At the top, the waves pass unhindered; but we note that waves also proceed in the shadow region as though coming from the knife edge itself. The knife edge acts as a new source, generating circularly cylindrical waves which propagate into the shadow area. In the area at the top of the figure the waves are not obstructed by the knife edge, and the waves continue to progress toward the right. The much weaker, cylindrical, diffracted waves are seen in the shadow area, progressing outward from a line located at the knife edge.

The comparable representation of the sound wave pattern existing in the

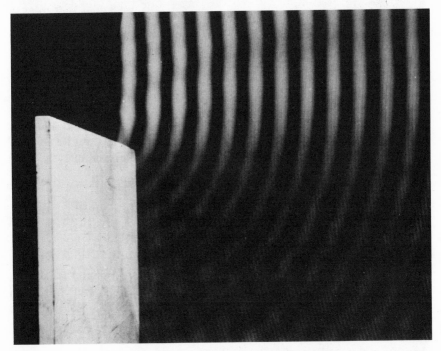

FIGURE 1.7. Plane sound waves arriving from the left proceed unhindered at the top of the figure. In the shadow region below, circular wave fronts caused by diffraction at the knife edge, are seen.

FIGURE 1.8. Sound waves diffracting around a circular disk combine in the shadow region to produce a central beam of parallel wave fronts.

shadow of an opaque circular disk is shown in Figure 1.8. Here the waves entering the shadow area originate at the perimeter of the disk, and they combine along the central line to form the "bright spot" of Lord Rayleigh.[4] Figure 1.9, made using the amplitude-only presentation technique, show the bright spot in the shadow region more clearly.

1.5. Diffraction by a Slit

First let us consider the case where the energy (and phase) is uniform over the entire face of the radiator. As shown in Figure 1.10, such a situation could result from the illumination of a slit located in an opaque screen by a distant light source or sound source. We are then interested in the radiation pattern existing in the dark area (shadow area) behind the screen. Figure 1.11 shows how this radiation (similar to the horn radiation of Figure 1.4) creates an illumination pattern on a second screen. We find that this pattern can be described by an expression known to radio and optical designers as a "sine

FIGURE 1.9. The amplitude pattern in the shadow of a disk shows the bright central lobe (the "bright spot" of Lord Rayleigh).

x over x" pattern. If we designate θ as the angle from the forward direction or axis, a as the aperture (the width of the slit), and λ as the wavelength, we find that the pattern amplitude A at a distant point is expressible as

$$A(\theta) = A_{\max} \sin x/x \qquad (1.2)$$

where $x = \pi a\theta/\lambda$.

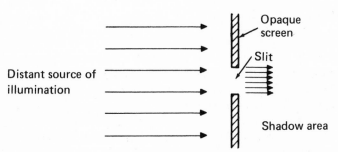

FIGURE 1.10. When a slit in an opaque structure is illuminated by a distant (plane wave) source of light, the energy distribution is uniform over the slit.

The exact expression is

$$A_{\max} = \frac{1 + \cos \theta}{2} \cdot \frac{\sin \left[(\pi a/\lambda) \sin \theta \right]}{(\pi a/\lambda) \sin \theta} \tag{1.3}$$

Equation (1.3) reduces to Equation (1.2) when θ is small.[5]

Equation (1.2) is plotted in Figure 1.12, and we see "side lobes" present. (Evidences of side lobes are present in Figure 1.4, and very prominent ones

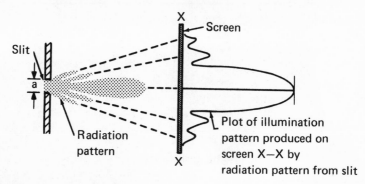

FIGURE 1.11. In the shadow behind the opaque screen of Figure 1.10, light emerges from the slit in a beam-shaped pattern.

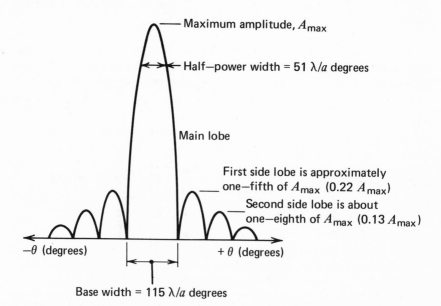

FIGURE 1.12. The diffraction pattern (beam pattern) formed by a slit.

show up in Figure 1.8.) We also see that the base width of the beam (in degrees) is given by $115\lambda/a$ and that the half-power width (in degrees) is $51\lambda/a$.

1.6. The Near (Fresnel), Far (Fraunhofer), and Transition Fields

We shall see later that in holography we are interested in diffraction effects produced nearby rather than at great distances. Near the slit in Figure 1.10 the light pattern remains the same width as the width of the slit, i.e., the beam remains completely collimated in this region.

We call the region which is far from the radiator the "Fraunhofer" region, after the Bavarian scientist Joseph von Fraunhofer. Near the radiator the pattern exhibits marked changes in its character, and the far-field or Fraunhofer pattern is generally considered as being adequately established only at distances exceeding $2a^2/\lambda$. For 100-wavelength aperture, the Fraunhofer region begins at a point 20,000 wavelengths from the radiator. For a 5-wavelength aperture, it begins at a point 50 wavelengths away. The field near the radiator is named after the French scientist Augustin Fresnel. It is usually considered to extend from the aperture to a distance equal to $a^2/(2\lambda)$, i.e., to a point one-fourth of the distance to where the Fraunhofer pattern starts. The region from $a^2/(2\lambda)$ to $2a^2/\lambda$, is often called the "transition region."

Figure 1.13 shows characteristics of certain field patterns produced by an illuminated slit in the Fresnel region and in the transition region. Very little energy "spreading" takes place within the Fresnel region, and the amplitude distribution is seen to shift gradually from its uniform value immediately in the vicinity of the radiator toward its final $(\sin x)/x$ pattern observed in the Fraunhofer region.[6]

Figure 1.14 illustrates sound waves being radiated from a lens placed in the mouth of a conical horn. Here the aperture is 30 wavelengths across. This figure gives only the amplitude of the sound waves, and, although it only shows the beam out to a distance of perhaps 30 or 40 wavelengths, it can be seen that the edges of the pattern remain parallel, i.e., the width of the beam remains the same as the aperture width for that distance.

1.7. Diffraction by Two Slits

We now consider the diffraction effects which occur when light passes through two slits in an opaque screen. Numerous very bright and very dark

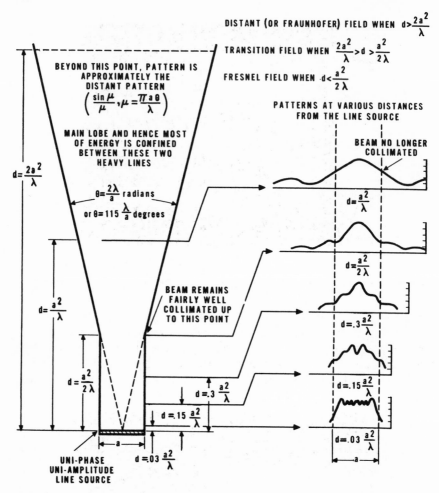

FIGURE 1.13. In the near field of the radiation emerging from the slit of Figure 1.11, a region exists where the beam remains collimated.

areas result, and these are called, in optical terminology, *fringes.* The two-slit effect of optics can be simulated with two nondirectional, coherent sound sources, as shown in Figure 1.15. This figure shows the fringe pattern formed by two sources separated by 3 wavelengths; in the figure the wavelength of the sound waves is approximately 1.5 in. The combination of two identical wave fields results in wave addition and wave cancellation. As would be expected, wave addition occurs at points equidistant from the two sources, i.e., along the center line of the two radiating points; this is the central, bright, horizontal area. Wave cancellation (destructive interference) occurs at those points where

FIGURE 1.14. By scanning the sound field in front of a 30-wavelength-aperture plane-wave sound source, the near-field variations can be accurately recorded.

the distance from one source differs from that to the other source by one half wavelength. At such points, one of the two wave sets has crests (positive pressures) where the other has troughs (negative pressures). The two areas where this half-wavelength destructive interference effect is evident in Figure 1.15 are the black areas immediately above and below the central bright area.

FIGURE 1.15. Two separated sound sources act like two optical slits in creating a diffraction (addition and subtraction) pattern (through constructive and destructive interference).

Bright areas are again seen above and below these two black areas. These are areas where the distances from the two sources differ by one *full* wavelength. One of the wave sets is a full wavelength ahead of the other and wave crests and wave troughs again coincide, so that positive pressures add to form higher crests and negative pressures add to form deeper troughs. Similarly, in those areas where one of the two wave sets is two wavelengths ahead of the other, wave addition again results; these are the shorter bright areas at the very top and very bottom of the figure.

1.8. Dependence of Diffraction on Wavelength

The diffraction pattern of Figure 1.15 was formed with single-frequency sound waves of a particular frequency (wavelength). Different-wavelength waves would have produced a different pattern. This effect is shown in Figure 1.16. Waves having the longer wavelength, λ_1, are one wavelength behind in the direction A, whereas those of wavelength λ_2 are one wavelength behind in a different direction.

1.9. Wave Polarization

Sound waves are longitudinal compressional waves. When generated in air by a vibrating object, the regions of lower and higher air density travel outward at the speed of sound (about 1100 ft/sec) as concentric spheres.

Electromagnetic waves (radio waves and light waves) are transverse waves, with the direction of the transverse motion being specified by their *polarization*. Radio waves can be generated by causing a rapid alternation in the direction of electric current flow in a conducting wire. The tall tower of a radio broadcasting station constitutes such a wire, and if a current of rapidly varying direction is caused to flow in the tower, electromagnetic waves are radiated in all directions. These are vertically polarized because the tower is vertical. For television broadcasting, shorter-wavelength radio waves are used, and, in the United States, horizontal polarization is utilized, with the television receiving antennas being horizontal (rather than vertical) rods. Because of the short wavelengths involved, highly efficient "dipole" receiving antennas need be only a few feet long. In England, vertical polarization is used for television, and there the forests of dipoles on city roof tops are vertical.

Since light waves are electromagnetic waves, they are also generated accomplished with devices using the Faraday effect. C. L. Hogan[8] used the Faraday effect in a ferrite inserted in a waveguide to achieve the first *microwave* nonreciprocal device, the *gyrator*. An acoustic gyrator was later described by the author, using a rotating tube and transverse sound waves.[9]

1.10. Refraction

Another phenomenon of importance is the deflection of waves entering a medium which has a different wave velocity. This is called refraction and can be visualized two ways. Figure 1.17 employs ray concepts, and Figure 1.18 utilizes the wave-front picture to show how the waves themselves are affected when they enter the lower velocity medium. The motion of wave fronts is often compared with the movement of rows of marching soldiers, since in both cases velocity changes bring about direction changes. It is apparent that Equation (1.1) (frequency times wavelength equals velocity) is involved in Figure 1.18. The frequency of a wave is a constant; hence a change in velocity must be accompanied by an equal change in wavelength, as indicated in Figure 1.18.

Because the ratio of the velocities of the two media provides a measure of the extent of the ray bending or refraction, it is called the index of refraction. The letter n is assigned to this quantity and the equation expressing the index of refraction of a particular substance is $n = v_0/v$, where v_0 is the free-space velocity and v is the velocity in that medium. Optically refractive substances such as water, glass, and diamond all have refractive indices which exceed unity, i.e., the wave velocity in these media is less than in free space. A diamond sparkles because it has a rather high index of refraction.

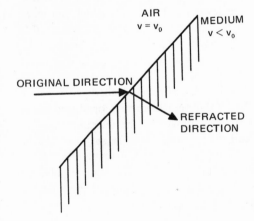

AIR
$v = v_0$

MEDIUM
$v < v_0$

ORIGINAL DIRECTION

REFRACTED DIRECTION

FIGURE 1.17. When wave energy leaves one medium and enters another obliquely, the ray direction is deflected if the velocities of propagation in the two media differ.

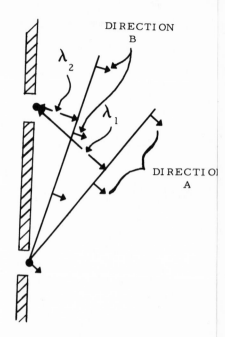

FIGURE 1.16. The diffraction angle for waves emerging from two slits, or many equally spaced grating slits, depends upon the wavelength. Because holograms are a form of grating, their reconstruction also depends upon wavelength.

electrically, usually by the rapid up-and-down or side-to-side motion o
myriads of the tiny, electrically charged particles (electrons). Because th
electrons can usually move in any direction, light waves from most ligh
sources are *randomly* polarized. Polarized light can be easily produced, how
ever, by the simple process of filtering out all waves except those polarize
in a given plane (as by the use of a sheet of Polaroid). This dichroic materia
absorbs components polarized parallel to a particular direction in the material
Certain other devices or materials can be used to affect the polarization o
electromagnetic waves. Thus, there are devices and materials which can *rotat*
the plane of polarization. Those which rotate by 90°, e.g., from a vertical t
horizontal direction, are referred to as *half-wave plates*. When the thicknes
is reduced by one-half, they become *quarter-wave plates*; these can caus
linearly polarized light to be transformed into circularly polarized light.[7]

Certain devices are *nonreciprocal*, i.e., the effect produced on polarize
light passing through in one direction is different from that produced in th
reverse direction. An example of such a device is the Faraday-rotatio
isolator, containing, for example, lead oxide glass. It can be set to rotate th
polarization by 45° during passage in one direction, and when the rotate
waves are reflected so as to pass back through the cell, their plane of polariza
tion is rotated *another* 45° in the same direction, resulting in the reflecte
waves having a 90° rotation relative to the incident waves. A similar result i

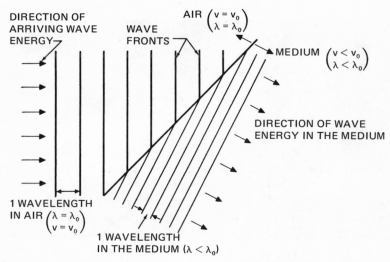

FIGURE 1.18. Wave energy entering a lower-velocity medium experiences a shortening of the wavelength, and the direction of propagation is altered.

1.11. Prisms

In Figures 1.17 and 1.18 we considered what occurred at the interface between air and a refractive medium (lower wave velocity). Let us now consider the case when wave energy enters and also emerges from a refracting substance. Figure 1.19 shows a cross section of a glass optical prism, with a light ray entering from the left side and emerging at the right. The ray is bent, or refracted, in entering the prism and it is again refracted upon leaving the prism. A wave-front portrayal of this phenomenon similar to that of Figure

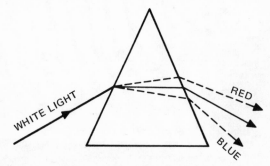

FIGURE 1.19. The wave velocity of red light entering a glass prism is not reduced as much as is the velocity of blue light.

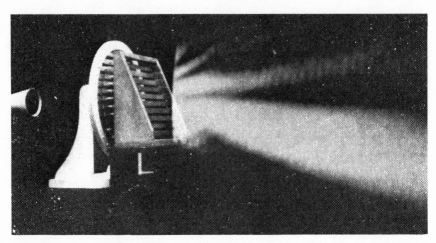

FIGURE 1.20. The focused beam of sound waves formed by an acoustic lens is deflected downward by an acoustic prism.

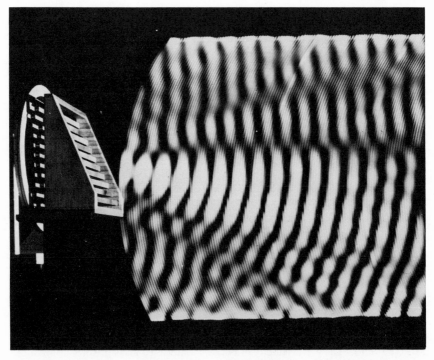

FIGURE 1.21. The wave pattern of the deflected sound waves of Figure 1.20.

1.18 would similarly indicate that the direction of the emerging ray (direction of wave propagation) would be as shown.

Prisms are often used to separate out the different wavelengths present in a composite electromagnetic wave. This is possible because many optical materials have refractive indices which vary with the wavelength or frequency of electromagnetic waves. As shown in Figure 1.19, a glass prism, which has a slightly different index of refraction for each of the various colors making up white light, deflects the red rays less than blue rays. It is thus able to delineate the various colors of the spectrum. A material whose refractive index is frequency dependent is called dispersive and a prism of such material is said to exhibit dispersion.

The deflection of wave direction by a prism can also be exhibited in the case of sound waves.[10] Figure 1.20 shows a situation where sound is first radiated by a small, conical, horn loudspeaker, then focused into a beam by a special acoustic lens, and then passed through an acoustic prism. Had the prism not been in place the beam would have been pointed in a horizontal direction; the prism has caused the beam to be deflected downward. Figure 1.21 shows the wave pattern of the deflected waves.

1.12. References

1. W. E. Kock and F. K. Harvey, Sound wave and microwave space patterns, *Bell Syst. Techn. J.* **20** (7), 564–587 (1951).
2. Photographing sound waves, *Bell Lab. Rec.* **July**, 304–306 (1950).
3. Lord Rayleigh, *Theory of Sound*, Dover, New York, 1945, Vol. 2, p. 142.
4. Lord Rayleigh, *Collected Papers*, Vol. 4, pp. 283, 305.
5. S. A. Schelkunoff, *Electromagnetic Waves*, Van Nostrand, Princeton, N.J., 1943.
6. W. E. Kock, Unpublished Bell Telephone Laboratories memorandum 44-160-204, Aug. 10, 1944.
7. W. E. Kock, Metal lens antennas, *Proc. IRE* **34**, 828 (1946).
8. C. L. Hogan, A microwave gyrator, *Bell Syst. Tech. J.* **31**, 1 (1952).
9. W. E. Kock, An acoustic gyrator, *Arch. Elek. Uebertr.* **7**, 106 (1953).
10. W. E. Kock and F. K. Harvey, Refracting sound waves, *J. Acoust. Soc. Am.* **21**, 471–481 (1949).

2

WAVE COHERENCE

In the preceding chapter the light waves discussed were referred to as single-wavelength or, what is the same thing, single-frequency waves. Waves which have this single-frequency property are said to have good frequency coherence. Light from a laser (particularly a gas laser) is said to exhibit an extremely high degree of frequency coherence. Because of the importance of wave coherence in holography, let us examine what the term coherence means.

2.1. Frequency Coherence

We noted that some water waves, some sound waves, and some light waves are single wavelength and rather uniform. One method of indicating the single-wavelength nature of a sound wave (the extent of its frequency coherence) is by portraying the *frequency analysis* of the waves. In Figure 2.1 such an analysis is presented for a single-frequency sound wave. The frequency content is plotted along the horizontal axis, and the amplitude or loudness of each frequency component is specified by the height in the vertical direction. The frequency or pitch of the one-frequency component of this sound is indicated as 440 Hz (the note *A* in the musical scale), and its loudness as two (arbitrary) units. Such a tone would be classed as having an extremely high frequency coherence.

An analysis of a somewhat more complicated sound is shown in Figure 2.2. This sound is one comprising a fundamental tone (the lowest frequency tone) and many overtones, with all overtones being harmonically related to

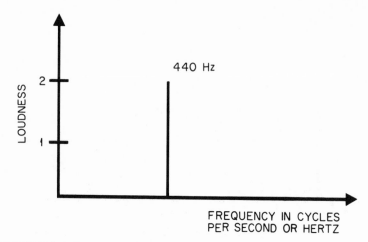

FIGURE 2.1. Representation of a wave comprising one single frequency.

the fundamental. Like the single-frequency sound, this tone also would be musical in sound because its waves constantly repeat themselves at the repetition rate of the fundamental.

A still more complicated sound is the sound of noise. Noiselike sounds are very irregular, and they therefore have very little coherence. They include, for example, the sound of a jet aircraft or the sound of howling wind in a

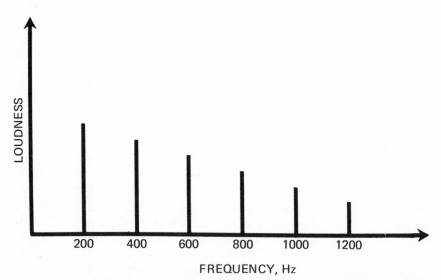

FIGURE 2.2. A periodic wave may have many harmonics, all integral multiples of the fundamental frequency.

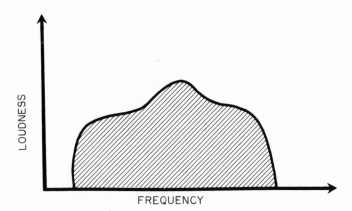

FIGURE 2.3. A nonperiodic wave such as the sound of noise can possess a broad spectrum of frequencies.

storm; such sounds have a broad, continuous spectrum of frequencies. The analysis of an exemplary noiselike sound is sketched in Figure 2.3.

The sound portrayed in Figure 2.1 is that of a perfectly pure, single-frequency sound; in other words, one of *perfect* frequency coherence. Actually, such absolutely perfect waves do not exist; however, the extent to which a wave *approaches* this perfection can be specified. Figure 2.4 shows the representation of a noiselike sound whose components extend over a fairly narrow frequency band. The extent to which such a signal approaches one having an infinitely narrow band, that is, a single-frequency content, can be specified by stating the width of its frequency band as a percentage of the center frequency. In Figure 2.4 the bandwidth is 10 Hz, and the central frequency component is 1000 Hz. The sound is thus one having a 1% frequency spread or a bandwidth of 1%.

2.2. Frequency Stability

Another way of specifying how closely a coherent wave approaches perfection is by stating its frequency constancy or frequency stability. The very fact that there is a frequency spread in the analysis of Figure 2.4 indicates that the frequency of the tone involved varies back and forth in pitch over the frequency band shown. It is accordingly not absolutely constant in pitch, and the specifying of a 1% bandwidth is equivalent to the statement that the tone has a frequency instability (or stability) of one part in one hundred.

Electronic sound-wave generators (called *oscillators*), such as those used in electronic organs, employ various procedures to achieve a high degree of pitch or frequency stability. Such procedures avoid the need for periodic tuning of the instrument. Some audio-oscillators achieve a frequency stability of one part in a million or better.

Radio waves are also usually generated electronically, and they too can be made very constant in frequency and, therefore, highly coherent. Figure 2.5 shows coherent microwaves issuing from a lens; the arrangement shown of Figure 2.6 was used in making that photograph (from Reference 1 of Chapter 1). The need for extremely high frequency-stable oscillators in certain radio applications led to the development of an exceedingly narrow-band radio device utilizing atomic processes to achieve its stability. It is called the *maser* (an acronym for *m*icrowave *a*mplification by *s*timulated *e*mission of *r*adiation). One variety, the hydrogen maser, can achieve a frequency constancy of one part in 10^{12}.

Following the development of the microwave maser, its principle was extended to the light-wave region, and light-wave "oscillators" having very high coherence then became available. These are now called *optical masers* or *lasers* (*l*ight *a*mplification rather than *m*icrowave *a*mplification).

One form of laser, the gas laser, exhibits a particularly high frequency

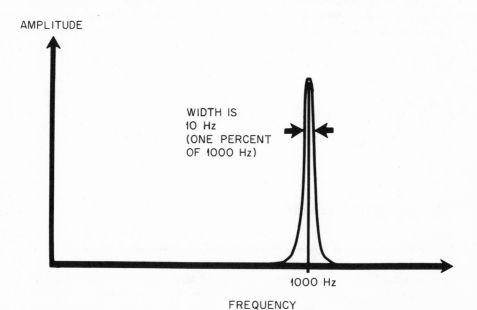

FIGURE 2.4. As the nature of a wave changes from nonperiodic to periodic, its spectrum narrows.

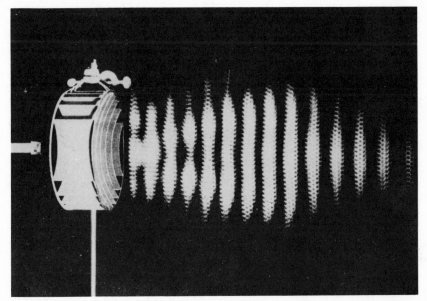

FIGURE 2.5. Spherical waves issuing from the waveguide at the left are converted into plane waves by a microwave lens. This photograph is a recording of the interference pattern formed by a wave set of interest and a reference wave; it can accordingly be considered a microwave hologram, with the vertical striations being microwave fringes.

constancy, and, accordingly, it is the one most often used for making holograms. Another form, the pulsed laser, provides higher-intensity light, but generally its light has less coherency. Earlier forms of coherent light sources (the ones available to Gabor in 1947) produce light which is less intense and less coherent than the light produced by a laser.

2.3. Spatial Coherence

With the laser, highly frequency-coherent light can be generated. For holography, however, a second form of coherence is required, namely, *spatial coherence*. Many of the figures of Chapter 1 showed single-frequency waves which were also very uniform.

The availability of such uniform waves is just as important in holography as the availability of single-*frequency* waves. This requirement is particularly evident when the uniformity of the reference beam is considered.

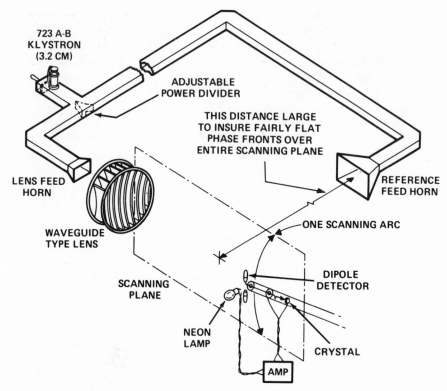

FIGURE 2.6. Reference waves from the feed horn interfere with waves emanating from the lens. Interference patterns are detected in the scanning plane by the dipole, amplified to light the lamp, and then photographed.

When plane light waves from a highly coherent laser are passed through a ground-glass plate, they are no longer plane waves; they are badly jumbled and mixed up. We shall see that, in reconstructing a hologram, the original reference beam and the later reconstructing beam must be alike; otherwise, the hologram scene will not be reproduced faithfully. If the original reference beam used were a badly jumbled one, the rather formidable problem would exist in the reconstruction process of providing a second, jumbled-up beam exactly like the reference beam.

2.4. Coherent Waves from Small Sources

Both spatial coherence and frequency coherence are needed for holography. Fortunately, once a wave generator is available which can generate waves having good frequency coherence, it is generally the case that the same generator can also form waves having good spatial coherence.

For sound waves or microwaves, spatial coherence is easily obtainable because the wave generators themselves can be made very small, only a wavelength or so across. It is seen, for example, in Figure 1.6 that the crest-to-crest distances, or wavelengths, of the sound waves portrayed are significantly larger than the size of the telephone receiver unit from which they are issuing. Also, in Figure 2.7, the very short radio waves called microwaves are seen issuing from the small, rectangular metal tube shown at the left. In microwave terminology this tube is called a *waveguide*, and, like the telephone receiver in Figure 2.6 it, too, is smaller in size than the 1.3-in.-wavelength microwaves issuing from it. Thus, both sound waves and microwaves can be radiated from sources whose dimensions are a wavelength or so in dimensions.

Figure 2.5 shows how the circular pattern of microwaves issuing from the waveguide at the left can be converted, by the structure immediately to its right (a metal microwave lens), into the plane-wave microwave pattern displayed at the right of the lens. The figure demonstrates that coherent microwaves, like coherent sound waves, can be placed in plane-wave patterns.

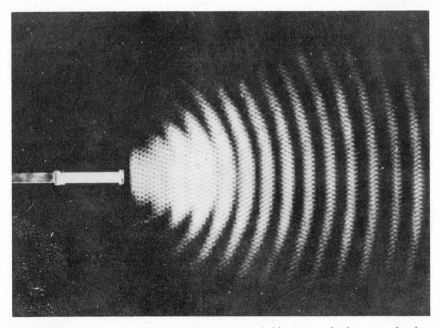

FIGURE 2.7. Here the lens of Figures 2.5 and 2.6 is removed; the waves issuing from the waveguide are seen to have circular wave fronts.

2.5. The Extremely Short Wavelengths of Light

The wavelengths of light waves are extremely small; thus, violet light has a wavelength of only 16 millionths of an inch (an optical lens 1.6 in. across would extend over 100,000 wavelengths of violet light). Because light wavelengths are so extremely small in size, it is practically impossible to construct a light generator the size of one light wavelength. Accordingly, the technique we have discussed for generating spatially coherent sound waves or radio waves, that of starting with a one-wavelength radiator and using lenses or reflectors to form spatially coherent plane-wave fields, is not feasible for generating spatially coherent light waves.

Thus, two problem areas exist in achieving coherent light; one is that of achieving good frequency coherence, and the other is that of achieving good spatial coherence. Ordinary light sources such as incandescent lamps are noiselike in nature, i.e., they exhibit a very poor degree of frequency coherence. Furthermore, they possess a luminous area which is extremely large compared to one light wavelength. Accordingly, even if the light from each and every tiny light generator on the incandescent surface were to be single-frequency light, spatial coherence would still be lacking because the light would be generated randomly and independently at millions of tiny points over the luminous area. This highly frequency-coherent, poorly spatially-coherent light would resemble laser light which has passed through a ground-glass screen.

We shall see that, fortunately, lasers differ from previous light sources in two ways. First, the light from a laser stems from one particular atomic energy-transfer process, and this causes its frequency content to be very pure (the light is single color or *monochromatic*). Secondly, it employs reflecting surfaces in the light-generation process which cause the light to be emitted in the form of extremely plane waves whose wave fronts are many, many wavelengths across. Lasers thus make available light having both good frequency coherence and good spatial coherence.

3

GRATINGS AND
ZONE PLATES

If the waves protrayed in Figure 1.15 had been single-wavelength light waves, rather than sound waves, and if the two radiators had been long slits perpendicular to the paper, a white screen placed to the right of these slits would have displayed a comparable series of bright and dark vertical bands, called, in optical terminology, *fringes*. We noted earlier, in connection with Figure 1.2, that when one of two wave sets is a full wavelength ahead of the other, the two again *add*, and, therefore, as in Figure 2.16, wave energy is diffracted into those off-angle directions for which such a condition holds.

3.1. The Optical Grating

Figure 1.15 portrays the diffraction pattern of *two* open slits; had there been many, equally spaced slits, wave energy would similarly have passed through the entire array of slits and would similarly have been diffracted into the same directions. Such an array of slits is called a *grating*. The undeviated (*zero-order*) waves manifest themselves as the horizontally directed white area in Figure 1.15. The *first-order* diffracted waves are those having, as in Figure 1.16, a one-wavelength slippage, and are seen as the first downward-tilted white area and the first upward-tilted white area. Also evident in Figure 1.15 are two still more widely deflected components (the smaller, more widely diverging white areas). These correspond to a slippage of *two* wavelengths, and are called the *second-order* diffracted waves.

A typical optical grating might be one having as many as a thousand narrow-ruled scratches made on a transparent glass plate, all within an inch or so. They might be scratched on the glass with a diamond, so that the scratches become opaque lines, with the clear glass sections between them constituting the open slits. The grating as we noted, splits a parallel beam of light into numerous packets that are deviated symmetrically to both sides of the direction of propagation of the original parallel beam. Walker[1] has demonstrated very clearly the diffraction of a beam of laser light into multiple orders by a (transmission) grating (Figure 3.1). In this figure the individual orders are made more apparent by the procedure of tapping two dusty chalk erasers against one another in the path of the diffracted rays. The narrow width of these rays shows the coherent (single-wavelength) nature of the laser light; as we saw in Figure 1.16, had the light in this figure comprised a band of wavelengths (e.g., one having a spectrum as in Figure 2.4), the various diffracted orders would have been much broader.

3.2. Slit Gratings and Photographic Gratings

In the multiple-slit grating we have been discussing, the regions between the open vertical slits are opaque. The wave amplitude behind these opaque areas is therefore very low, wheras that behind the open slits is high, as shown in the top portion of Figure 3.2.

A grating can also be made photographically. This procedure is shown in

FIGURE 3.1. Chalk dust displays the rays of laser light diffracted by a transmission grating (courtesy of D. K. Walker).

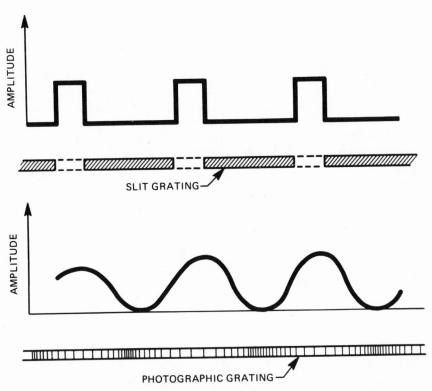

FIGURE 3.2. The classical slit grating (top) has abrupt changes in light transmission, whereas the photographic grating (below) does not.

Figure 3.3. Plane coherent waves propagating in a horizontal direction (indicated as wave set *A*) are made to interact with another set of plane waves having the same (single) frequency and arriving at any angle (set *B*). The interference pattern generated at the plane of a photographic plate is then photographically recorded as shown. The combination of sets *A* and *B* causes wave *addition* to occur along these horizontal lines of the photographic plate where the positive crests of the two sets reinforce each other (marked + +). Wave diminution occurs where the positive crest of one meets the negative trough of the other (marked + −). The light intensity is greater along these lines where the light energy adds, and accordingly, the plate is more strongly exposed there. Conversely, along these lines where a diminution of energy exists, the plate is more weakly exposed. Parallel striations of light are thus recorded on the photographic film, and after the plate is developed and fixed, these striations appear on the film as lines. A photographic record of this type is shown in Figure 3.4. Such a photographically made grating differs from the simple multiple-slit grating, as shown in the lower part of Figure 3.2. In the photographic recording, the change from the transparent to the opaque area is gradual, rather than sudden. This causes the light amplitude to vary *gradually* (rather than suddenly) from its maximum to its minimum value. This smooth variation in light intensity causes the bulk of the wave energy to be diffracted into only the zero-order component and the two first-order components, rather than into many orders, as in Figure 3.1. This process of diffraction by a photographically made grating is shown in Figure 3.5. The photographic plate is drawn so as to resemble the one in Figure 3.3; the zero-order waves are undeviated, and the two first-order components are labeled upward and downward. It is pertinent to note here that the *direction* of the downward-deviated wave is identical to the direction of the tilted wave set of Figure 3.3. which was used to form the photographic grating. The photographic grating also differs from the simple slit grating in that the process of developing and fixing exposed film removes those light-sensitive elements still remaining unexposed in the emulsion, and the process causes the emulsion to exhibit varying amounts of shrinkage over the area of the photographic plate. The amount of this shrinkage, though quite small, can be significant at light wavelengths.

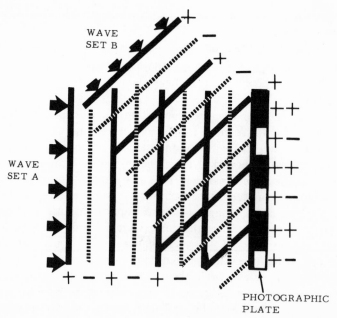

FIGURE 3.3. When two sets of single-wavelength plane waves meet, interference occurs; at the photographic plate, addition and cancellation occurs.

FIGURE 3.4. At the plane of the photographic plate of Figure 3.3., the wave interference pattern is a series of bright and dark lines. This is a photographic recording of such a pattern.

3.3. A Prism Grating

Because emulsion shrinkage causes photographically made gratings to have varying thicknesses, wave-*tilting* (refraction) effects can occur for them. Consider a grating comprised of many prism sections, instead of many slits, as shown in Figure 3.6. Each prism section tilts the wave direction downward, as does the single prism of Figure 1.20 and 1.21. If, in addition, the light wavelength and the vertical prism spacing are properly selected, the light diffracted into the first-order downward wave direction can be greatly enhanced by the prism. If the spacing and wavelength are chosen so that waves moving in the direction of the prism tilt are also one wavelength behind (or ahead) of the waves issuing from their neighboring prism, the two effects add, and practically all of the energy passing through is deflected in the direction indicated by the arrows in Figure 3.6.

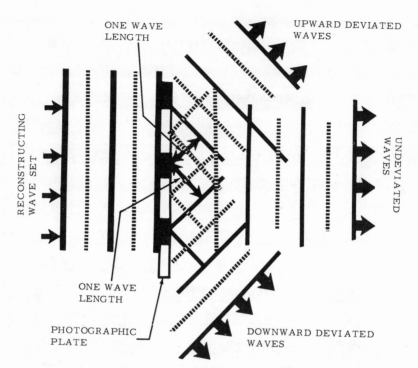

ONE WAVE LENGTH

UPWARD DEVIATED WAVES

RECONSTRUCTING WAVE SET

UNDEVIATED WAVES

ONE WAVE LENGTH

PHOTOGRAPHIC PLATE

DOWNWARD DEVIATED WAVES

FIGURE 3.5. When the line pattern of Figure 3.3 (or 3.4) is illuminated with the original *horizontally* traveling set of plane waves, one of the two off-angle diffracted waves travels in the same direction as that of the original off-axis wave set (of Figure 3.3).

Similar diffraction–refraction enhancement procedures have been employed in the design of some microwave lens antennas.[2] There, the process is called *stepping*, or *zoning*. This procedure can be used in microwave applications because the wavelengths employed often cover only a rather narrow band of frequencies, and they can, therefore, be treated as single-wavelength devices. A stepped, or zoned, circular, microwave lens, used in the first microwave relaying of television programs between New York and Boston by the Bell System, is shown in Figure 3.7.

3.4. A Double-Prism Grating

An extension of the prism-grating concept of Figure 3.6 permits enhancement of both upward and downward first-order waves: this concept is illustrated in Figure 3.8. Here double prisms are used, with the top halves acting as downward-tilting prisms, and the lower halves causing wave energy to be tilted upward.[3] Again, it is assumed that the vertical spacing and the wavelength are correctly chosen to cause the prism-tilt directions to correspond to the first-order diffraction directions.

If we assume that, after developing and fixing the photographically recorded grating shown at the bottom of Figure 3.2, some emulsion shrinkage has occurred, we see that the top and bottom horizontal lines of grating will become wavy lines. Ridges and troughs will have been created, with the troughs positioned at the areas of maximum shrinkage. We have already noted that even a small amount of shrinkage can be significant because of the short wavelength of light. Of importance, also, is the fact that the *spacing* of these emulsion prisms is exactly right to bring about the effects described.

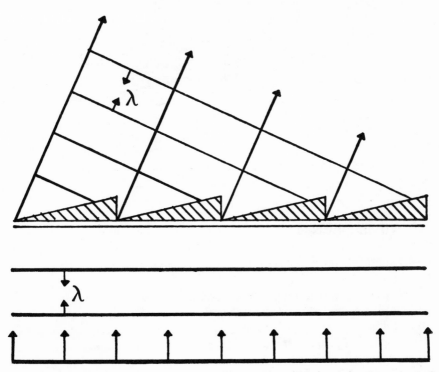

FIGURE 3.6. An array of prisms can act as a very efficient grating for waves of the proper wavelength.

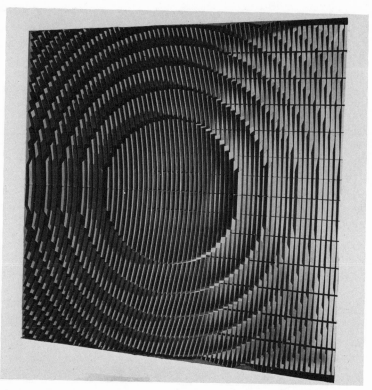

FIGURE 3.7. A microwave lens, designed in 1946 and employed in the Bell System's New York-to-Boston microwave radio relay circuit, has a circular "step" design which matches that of a zone plate. The relation of this lens to a computer-generated hologram, called a kinoform, will be discussed later.

3.5. Volume Effects

A third difference between the simple slit grating and the photographic grating arises from the fact that the photographic emulsion is many light-wavelengths thick. The interference pattern generated by two sets of light waves is a three-dimensional one. When this pattern is recorded photographically, it does not merely exist at the *surface* of the photographic film (as was tacitly assumed in connection with Figure 3.3); it establishes itself throughout the three-dimensional *volume* of the film emulsion. Because the film emulsion has an appreciable thickness, this volume pattern, as recorded in the photographic film, can be quite significant. Photographically made gratings, zone plates, and holograms must therefore be considered as recorded volume interference patterns. We shall discuss these volume effects in more detail shortly.

3.6. The Classical Zone Plate

The zone plate can be described as a set of flat, concentric, annular rings which diffract wave energy. The open spaces permit passage of waves which add constructively at a desired focal point, and the opaque rings prevent passage of waves which would interfere destructively with these waves at that point.[4]

Figure 3.9 indicates the procedure for determining the postions of the rings in a zone plate having a circular opaque disk for its central portion. We recall from Figures 1.8 and 1.9 that an opaque disk can effect a concentration of energy along its axis by virtue of wave energy being diffracted by it into the shadow region. At some point along this axis a further concentration of energy can be provided by the use of opaque rings which allow only that wave energy to pass which will interfere constructively at that particular point on the axis. The proper positions of the blocking rings are determined, as shown in Figure 3.9, by drawing circles whose centers coincide with the desired focal point and whose successive radii from this focal point differ from one another by one-half of the design wavelength. Thus, at that point where the first half-wavelength circle intersects the plane of the zone plate, the central blocking zone, the opaque disk is terminated. Farther out, at the

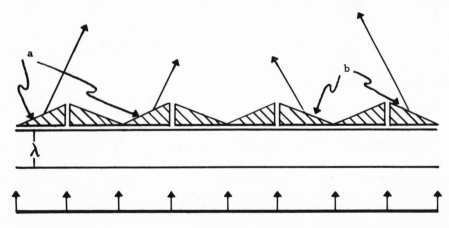

FIGURE 3.8. An array of double prisms, acting as a grating, can cause both first-order diffracted waves to be reinforced. Emulsion shrinkage in a photographically recorded grating can introduce thickness effects comparable to this double-prism grating.

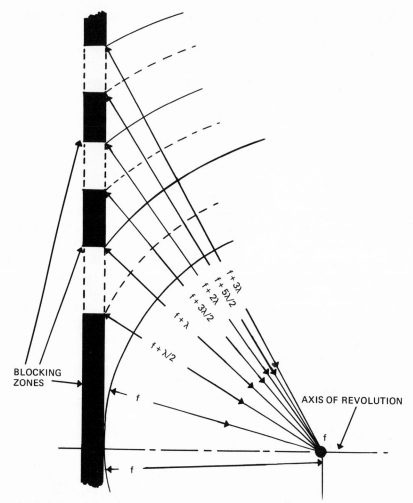

FIGURE 3.9. Design of a zone plate having an opaque disk as its center element.

1-wavelength point, the first annular blocking ring is started. It is terminated at the 3/2-wavelength point, and the process continues. It should be noted that the design of a zone plate is based on one particular wavelength; waves having wavelengths which differ from this design wavelength will not be affected by it in the desired way. A frontal view of such a zone plate is shown in Figure 3.10.

Zone plates can be used at light wavelengths; they can be used for concentrating microwaves; and they can be used in acoustics. Lord Rayleigh noted that a zone plate designed for sound waves can play "the part of condensing lens." An antenna of a World War II microwave radar incorporated a zone plate,[5] and some microwave lens antennas now in use employ the zone concept.[6] In these uses the microwave lens thickness is altered at the 1-wavelength zone positions; Figure 3.7 is an example. As in the case of the prism grating of Figure 3.6 compared to a slit grating, zoned *lenses* are more efficient than zone *plates* because they avoid the discarding of that energy which is reflected or absorbed by the blocking zones.

3.7. Zone Plates with Areas Interchanged

One property of zone plates that is pertinent to holography is that their functioning is unaffected if the blocking zones and open-area zones are interchanged. This is evident from Figure 3.11. There the central and other blocking zones in Figure 3.9 are made open-area zones, and the waves passing through them still differ from one another at the focal point by an integral

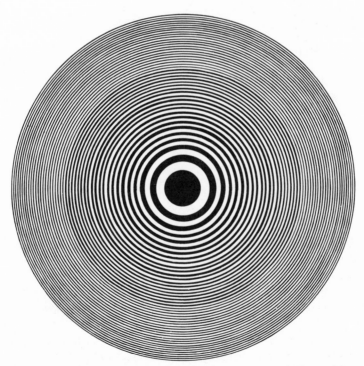

FIGURE 3.10. Head-on view of the zone plate of Figure 3.9.

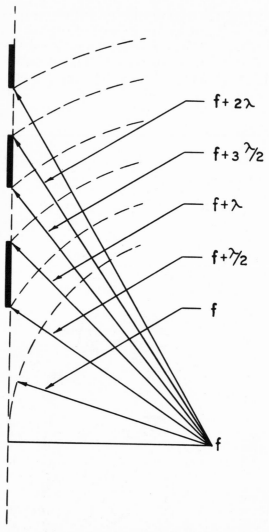

FIGURE 3.11. The opaque blocking zones of Figure 3.9 can be replaced with open spaces and the zone plate will still function as before.

number of wavelengths. Additionally, since the originally open areas (now blocking areas) differ by a half wavelength from the new open areas, the desired zone plate action is unaffected.

This ability of a zone plate to function properly with transparent and opaque zones interchanged can be even further extended; the size of the center area, whether blocking or open, can be varied. It is then only necessary to make the next (surrounding) zone area of such a size that the distance from the desired focal point to its outer edge is a half wavelength greater than the length of its inner edge. The next area is then made a half wavelength greater than that, etc. Again, full zone-plate action results since additive waves are passed and cancelling waves are blocked. This interchangeability property of zone plates permits them to be easily recorded photographically.

3.8. A Photographic Zone Plate

The photographic recording of the interference pattern between plane and spherical single-frequency light waves is illustrated in Figure 3.12. It turns out that the spacings of the photographic rings which are generated in Figure 3.12 are identical to those of a zone plate. An actual record of this kind is shown in Figure 3.13. Because sections of this pattern located near the central top and central bottom edges of this pattern resemble somewhat (except for the curvature) the horizontal line pattern of an optical grating (Figure 3.4), we would again expect three components to be generated when this developed pattern is again illuminated with plane waves.

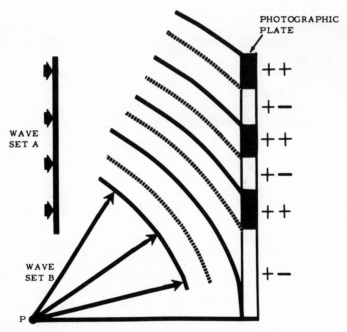

FIGURE 3.12. Plane and spherical waves can be combined to form a *photographically* made zone plate.

FIGURE 3.13. The photographically recorded interference pattern generated by combining coherent plane and spherical wave is a zone-plate pattern.

This effect is shown in Figure 3.14. A portion of the plane waves arriving from the left is undeviated, passing straight through the photographic transparency. Because the circular striations act like the lines of a grating, energy is diffracted both upward and downward. However, the pattern of striations is circular, and the waves which are diffracted in the upward direction travel outward as circular wave fronts, as though originating at the point P_v. These waves form what is called a *virtual image* of the original point light source P of Figure 3.12 (virtual, because in this reconstruction no source really exists there). These waves provide an observer, located where the words "upward waves" appear in Figure 3.14, with the illusion that an actual point

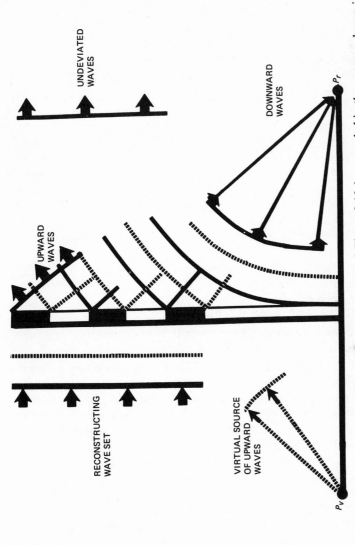

FIGURE 3.14. When the upper portion of the circular line pattern of Figure 3.13 (as recorded in the process shown in Figure 3.12) is illuminated with the original, horizontally traveling set of plane waves, three sets of waves result. One set travels straight through horizontally; another acts as though it were diverging from the source point of the original spherical waves; and the third is a set which converges toward a point on the opposite side of the recorded circular pattern.

source of light exists there, fixed in space behind the photographic plate no matter how he moves his head. Furthermore, this imagined source exists at exactly the spot occupied by the original spherical-wave light source used in making the photographic record.

A third set of waves is also formed by the recorded interference pattern. In Figure 3.14 this set is shown moving downward, and these waves are focused waves. They converge at a point which is located at the same distance to the far side of the photographic record as the virtual source is on the near side. The circular striations of the photographic pattern cause a *real image* of the original light source P to form at P_r (*real*, because a card placed there would show the presence of a true concentration of light).

The *completely* opaque and *completely* transparent sections of the zone plates of Figures 3.9 and 3.11 are not observed when the zone plate is made photographically. The difference lies in the fact that when the two wave sets interfere, the variation from complete destructive interference to complete constructive interference occurs in a gradual manner. This effect was illustrated in Figure 3.2, where the different transparencies of a group of zones of a slit grating and those of a photographically made grating was illustrated. In the photographic recording, the change from transparent to opaque area is gradual, and, when illuminated, the transmitted light strength varies gradually from a maximum to a minimum value. As in the case of the photographic grating, this sinusoidal variation in light amplitude causes the major portion of the incident light to be diffracted into the two first-order directions (and into the zero-order component). For optical zone plates, this simultaneous diverging and converging property results in multiple-image effects; when a zone plate is placed in front of an object, several images are seen. This effect is shown in Figure 3.15. The zone plate used in the figure was a very precisely made photographic recording of the interference pattern, formed by light waves from a point source interfering with plane waves,[7] as shown in Figure 3.16. In Figure 3.15, the zone plate is the darker circular shadow area. The zone plate is imaging, off-axis, an object which looks like a square, white, picture frame. Of the three "images" seen, the object itself is the brightest and clearest because the zone plate is fairly transparent; thus "image" is what we have called the zero-order, undeviated, component. To either side of it is another, slightly blurred, image of the white frame. The one deflected toward the center of the dark circular area is produced by the converging waves, and the other is formed by the diverging waves. It is worthy of note that here only the zero-order and the two first-order images are at all prominent. This would not be the case if a classical, blocking zone plate were involved. The gradual change from transparency to opacity in the fringes of the photographic (hologram) zone plate causes the higher order images to be suppressed. Complete separation of images does not occur in Figure 3.15 because the object

FIGURE 3.15. Double-image effects are produced by the converging waves generated by a zone plate. For this figure a 2-in.-diameter zone plate was used, made, as indicated in Figure 3.16, by causing spherical waves and plane waves to interfere on a photographic plate. The object to be imaged is the white square shown as the central object. When the zone plate is moved to an off-axis position, a partial separation of the three components results. One of the two fainter images is produced by the zone plate's positive lens action, the other by negative lens action. If the zone plate had been offset still further, the three diffracted components would have been completely separated.

is too near the axis of the zone plate. Had it been placed further off-axis, all three images would have been fully separated and would appear as separate, individual images.

A zone plate made photographically, as in Figure 3.16, has an intensity distribution $I(x, y)$ at the plane of the photographic plate given by

$$I(x, y) = \{C_1 + C_2 \cos [a(x^2 + y^2)]\}^2 \tag{3.1}$$

This sinusoidal variation causes two focal points to exist,

$$f = \pm \, \pi/(a\lambda) \tag{3.2}$$

where λ is the wavelength illuminating the zone plate. The $\pm$ sign denotes the fact that a focal point on one side receives converging energy (as a convex lens focuses energy) and the (conjugate) focal point on the opposite side acts as a source of diverging energy (as from a concave lens).

When the zone plate is made mechanically, such as by a ruling machine,

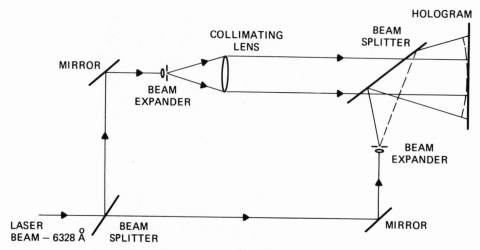

FIGURE 3.16. The procedure used to generate, very accurately, the photographic zone plate of Figure 3.15 for later use as a rather high-quality optical "lens."

the abrupt changes at a sharp boundary cause higher-order foci to be generated, these corresponding to powers which are odd-integer multiples of the primary power. These higher-order foci are analogous to the large number of Fourier coefficients associated with a square-topped periodic wave.

3.9. Offset Zone Plates

To clarify the action of an offset zone plate, let us review certain characteristics of two focusing devices, the parabola and the zone plate. Parabolic reflectors are often used as microwave antennas. The ability of such a reflector to form a beam of plane waves is traceable to a geometrical property of a parabola. Very often, only portions of such reflectors are used. Several of the antennas shown in Figure 3.17 are such devices; only a part of the parabolic (paraboloidal) surface is used to reflect incoming energy into the horn. Such offset reflectors are now used widely in microwave relays (Figure 3.17 is a tower on such a relay circuit).

Similarly, an offset section of a zone plate can focus wave energy. This procedure is illustrated in Figure 3.18. Plane waves arriving from the left impinge upon a portion of a zone plate, causing the energy to converge on the focal point f. In addition, this portion will, through its negative lens action, generate diverging waves appearing to emanate from the conjugate focal point f_c.

FIGURE 3.17. Present-day radio relay circuits utilize horn-lens antennas (the square unit) and offset, horn-enclosed, parabolic reflectors (A.T.&T. photograph).

3.10. Zone Plates as Lenses

Photographically made zone plates have been suggested for use as lightweight lenses in the fields of communication, astronomy, and space.[7] It was suggested that they be photographically recorded on lightweight plastic sheets which occupy a very small space and can later be "unfurled" to their full aperture size for concentrating laser communication beams or for taking photographs. Figure 3.19 shows the zone plate, which was used in making Figure 3.15, causing sunlight to be focused on a white sheet of paper.[8] That photographically made zone plates can act as camera lenses is shown in Figures 3.20 and 3.21. These figures were made with the camera of Figure 3.22, having the photographically made zone plate used in making Figure 3.15 as its only lens.[8]

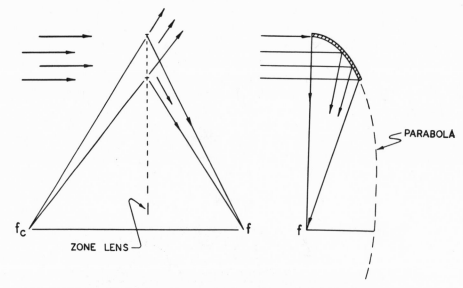

FIGURE 3.18. (Right) An off-center section of a paraboloidal microwave antenna such as the ones in Figure 3.17 concentrates wave energy at its focal point (*f*). (Left) An off-center section of a zone plate causes some wave energy to be focused and some to diverge from a conjugate focal point (f_c).

3.11. Volume Zone Plates

Zone plates used for microwaves and sound waves and the ruled-glass zone plates used for light wavelengths are generally considered to be two-dimensional (planar) structures. Because of the extremely short wavelengths of light waves, photographically made zone plates exhibit three-dimensional, or volume, effects, and this effect must be taken into account for them. The emulsion in a photographic plate may be many light wavelengths thick, so that *three-dimensional* interference patterns become recorded in the emulsion.

Let us examine these photographically recorded volume effects for zone plates, because these same effects also provide the basis for reflection holograms and three-color holograms.

3.12. Standing-Wave Patterns

As noted, interference effects exist not only at the surface of the photographic plate, but throughout the volume of the emulsion, so that if the photographic emulsion is many wavelengths thick, it will be exposed in depth and will record a volume interference pattern. The possibility of recording three-dimensional (volume) light-wave patterns within the emulsion of a photographic film was recognized in the early 1900's by the French scientist G. Lippmann.[9] He proposed using such recorded patterns for a form of color photography. His patterns were *standing-wave* patterns, and because these are also of importance in certain holograms and zone plates, let us review the characteristics of standing waves.

A standing-wave pattern can be looked upon as being made up of two waves, one moving to the left, and the other moving to the right. Standing waves can thus be considered as being formed either by a reflection of wave motion at a wall, or, by two single-frequency waves moving in opposite

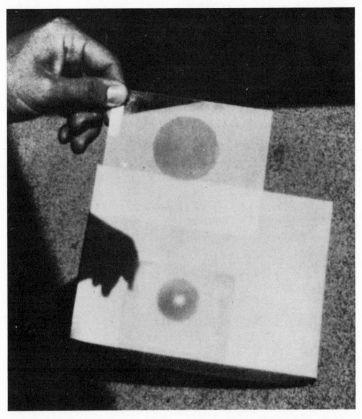

FIGURE 3.19. The (hand-held) 2-in. hologram zone plate of Figure 3.15 can cause sunlight to be focused on a sheet of white paper below. The dark circular area is the zone plate.

FIGURE 3.20. A photograph of three bars taken with a camera using the 2-in.-diameter zone plate of Figure 3.15 as its only lens.

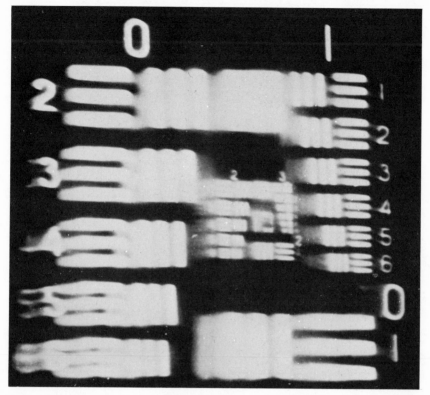

FIGURE 3.21. A photograph of a section of a standard optical test chart, also taken with the 2-in.-diameter zone plate as the only lens.

directions. It is this latter, oppositely moving wave case that is of interest to us in holography and zone plates.

In Figure 3.23, plane-wave laser light is shown arriving from the top of the figure, and it is split into two horizontal beams by mirrors A and B. By means of the other four mirrors, the two beams are caused to be superimposed and to interfere with each other while traveling in opposite directions. They thereby generated a standing-wave pattern, and in the figure this is shown as being formed within the emulsion of a photographic plate. It is evident that there will be regions (planes) where destructive interference (low light intensity) exist; in these areas the emulsion will be very lightly exposed. At the in-between, *strong* light intensity planes, the emulsion will become more heavily exposed. When the emulsion is developed and fixed, it will exhibit heavily exposed vertical planes spaced apart one half wavelength of the original laser light.

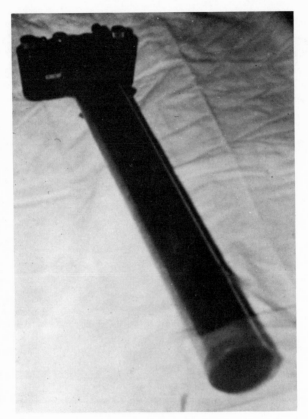

FIGURE 3.22. The long-focal-length lens (zone plate) of Figure 3.15 used as the only lens in a camera.

3.13. Lattice Reflectors

The half-wavelength lattice structure just described is known to be highly reflective for waves whose half wavelengths correspond to the lattice separation. This reflection phenomenon can be explained with the aid of Figure 3.24. Here waves arriving from the right are reflected successively by the numerous, rather densely exposed (but not opaque), planes *1, 2, 3, 4* located within the photographic emulsion. Some waves, *A*, are reflected at the first surface *1*. Other wave energy which passes through the various surfaces is indicated as waves *B*, *C*, and *D*. Waves *B*, when reflected at plane *2*, rejoin waves *A* at a distance 2*a* behind them. Similarly, waves *C* are a distance 4*a* behind them, and waves *D*, a distance 6*a*. If *a* is equal to one half wavelength of the incident light, all three of these additional distances are exact multiples of one wavelength, and all reflected wave sets *A*, *B*, *C*, and *D* will be in perfect step; the crests of all will coincide, as will the troughs. Thus, for light of wavelength 2*a*, reflections from all planes will add constructively, and the *reflectivity* of this structure to that particular wavelength light will be very high.

Because other-wavelength light will not add in this way, the reflectivity for them will be significantly lower. Accordingly, when this multiple structure is illuminated with white light, which comprises all the colors (wavelengths) of visible light, it will reflect only a single-wavelength (single-color) light. If, in the developing and fixing of the photographic plate, no shrinkage has been permitted (something which is not easy to accomplish), the color of this reflected light will be the same as that of the laser light which was responsible for forming it.

This procedure is the basis of the Lippmann color process; as noted, it was recently used to permit holograms to be viewed with white light instead of laser light (these are called reflection holograms).[10,11] The process was later extended to making *color* reflection holograms, exposing the hologram plate successively with two or three differently colored laser wave sets.

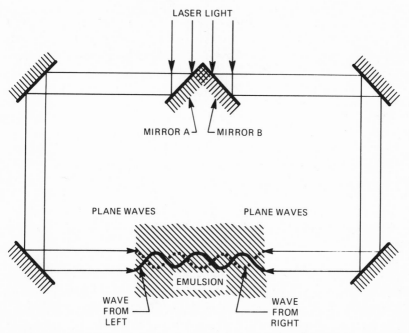

FIGURE 3.23. Coherent waves traveling in opposite directions generate standing waves. In a photographic emulsion, these appear as densely exposed and lightly exposed planes.

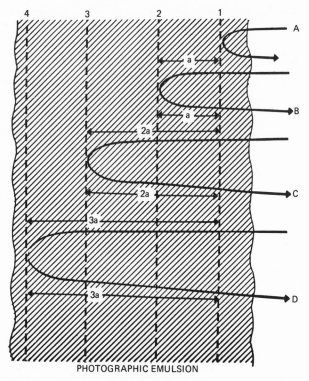

FIGURE 3.24. The reflecting planes of Figure 3.23 cause waves having a wavelength equal to twice the plane spacing to be reflected as constructively interfering waves.

3.14. Reflection Zone Plates

Zone plates can also be made using this three-color reflection procedure. Plane waves and spherical waves are made to approach the photographic plate from opposite sides, as shown in Figure 3.25. Their interference generates a zoned, longitudinal, lattice structure within the emulsion. When the emulsion is developed, fixed, and illuminated with white light, reflection and focusing of only one color occurs, the remaining colors passing on through. If this reflection zone plate is used as a lens or focusing device, a monochromatic image is formed.

To make a *three*-color zone plate (a zone-plate camera lens), the photographic plate would be exposed, as shown in Fig. 3.25, not once, but three times, using three differently colored laser sources successively.[8] The de-

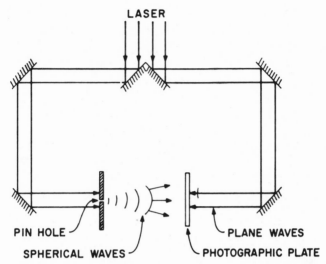

FIGURE 3.25. By causing coherent plane and spherical waves to impinge on a photographic plate from opposite sides, a reflection zone plate is produced. We shall see that the process is similar to that used in forming reflection (white-light) holograms.

veloped emulsion would then contain three longitudinal lattices, which, when exposed to white light, would reflect and focus only those colors of the white light having wavelengths corresponding to the three different lattice spacings. If this reflection lens were to be used as the lens in a camera using standard color film, it would properly image three differently colored views of the scene on the film, and satisfactory color reproduction of objects or scenes would result.

We have discussed zone plates at some length because, as we shall see in the next chapter, their diffraction properties are quite similar to those of holograms, a relationship recognized in 1950 by the British scientist G. L. Rogers.[12]

3.15. References

1. D. K. Walker, Visible diffracted rays, *Phys. Teach.* **11**, 435 (1973).
2. W. E. Kock, Radio lenses, *Bell Lab. Rec.* **May**, 193–196 (1946).
3. W. E. Kock, *Lasers and Holography*, Doubleday, Garden City, N.Y., 1969, p. 54.
4. F. A. Jenkins and H. E. White, *Fundamentals of Optics*, McGraw-Hill, New York, 1957, p. 360.

5. E. Bruce, Directive radio system, U.S. patent 2,412,202, Dec. 10, 1946.
6. W. E. Kock, Metal lens focusses microwaves, *West. Electr. Oscill.* May, 18–20 (1946).
7. W. E. Kock, L. Rosen, and J. Rendeiro, Holograms and zone plates, *Proc. IEEE* 54 (11), 1599–1601 (1966).
8. W. E. Kock, Three-color hologram zone plates, *Proc. IEEE* 54 (11), 1610–1612 (1966).
9. G. Lippmann, Sur la theorie de la photographie des couleurs simples et composes par la methode interferentielle *J. Phys. (Paris)* 3, 97 (1894).
10. Y. N. Denisyuk, *Dokl. Akad. Nauk SSSR* 144, 368 (1962).
11. G. W. Stroke and A. Labeyrie, White light reconstruction of holographic images, using the Lippmann–Bragg diffraction effect, *Phys. Lett.* 20, 368–370 (1966).
12. G. L. Rogers, Gabor diffraction microscopy: the hologram as a generalized zone plate, *Nature (London)* 166, 237 (1950).

4

HOLOGRAM
FUNDAMENTALS

In forming a hologram, two beams from the same laser are made to interfere. One beam is the light reflected from the scene to be photographically recorded; almost invariably, it is an extremely complicated one. The other is usually rather simple, often being a set of plane waves. This second set is called the reference wave, and, in reproducing or reconstructing for the viewer the originally recorded scene, a similar set is used to illuminate the developed photographic plate, the hologram.

The two original sets of hologram waves are caused to interfere at the photographic plate, as shown in Figure 4.1. Here the "scene" comprises a pyramid and a sphere. The objects are illuminated by the same source of single-wavelength laser light which is forming the plane waves at the top of the figure. Because the wave fronts of the set of waves issuing from the scene are quite irregular, the interference pattern in this case is quite complicated. After exposure, the photographic plate is developed and fixed, and it thereby becomes the hologram. When it is illuminated with the same laser light used earlier as the reference wave, as shown in Figure 4.2, a viewer imagines he sees the original two objects of Figure 4.1 in full three dimensions.

4.1. Two Simple Holograms

To understand how such a slight-wave interference pattern, once photographically recorded and then developed, can later re-create a lifelike image of the original scene, let us recall the very simple interference pattern formed

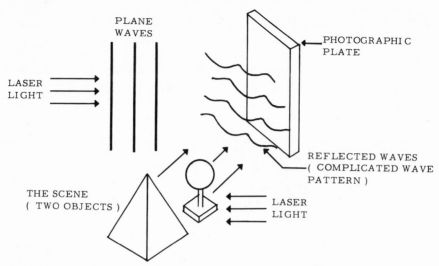

FIGURE 4.1. In making a hologram, the scene is illuminated with laser light, and the reflected light is recorded, along with a reference wave from the same laser, on a photographic plate. The plate is then developed and fixed.

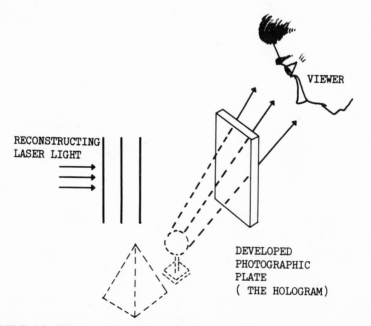

FIGURE 4.2. When the developed plate of Figure 4.1 is illuminated with the same laser reference beam, a viewer sees the original scene reconstructed, standing out in space behind the hologram "window."

when two sets of plane waves were made to interfere as in Figure 3.3. When this pattern was recorded, it became a photographic grating, and when the grating was illuminated with the same horizontally propagating plane waves used in forming the interference pattern, one of the diffracted components (the one propagating in the downward direction) proceeded in exactly the same direction that wave set *B* of Figure 3.5 would have traveled had the photographic plate not been present. Accordingly, a viewer in the path of these *reconstructed* waves would imagine that the source which generated the original set *B* in Figure 3.3. was still located behind the hologram. The photographic *hologram grating* is thus able to "regenerate" or reconstruct a wave progression long after it has ceased to exist. The hologram grating also generates a second set of upward-moving waves which were not present originally. We shall see that all holograms "reconstruct" *two* sets of waves.

Let us consider next the photographically made zone plate discussed in connection with Figure 3.12 and 3.14. Here again two deviated components were formed when plane waves illuminated the recorded zone plate, and the upward-moving waves were diverging, appearing to come from the point P_v, which was the location of the point source used in forming the interference pattern. These waves form what is called a *virtual image* of the original point light source *P* (virtual, because in the reconstruction, no source really exists there). These waves give to an observer, located where the words "upward waves" appear in Figure 3.14, the illusion that an actual point source of light exists there, fixed in space behind the photographic plate, no matter how he moves his head. Furthermore, this imagined source exists at exactly the spot occupied by the original spherical-wave light source in making the photographic record. In Figure 3.14, there is a downward-moving component, and because the recorded pattern is circular, these waves are focused waves. They converge at a point which is located at the same distance to the far side of the photographic record as the virtual source is on the near side. The circular striations cause a *real image* of the original light source *P* to form at P_r (*real*, because a card placed there would show the presence of a true concentration of light).

4.2. The Complete Hologram Process

The complete, two-step, zone-plate, hologram process of Figures 3.12 and 3.14 is shown in Figure 4.3. Here, a pinhole in the opaque card at the left serves as the "scene;" it is a point source of spherical waves. These interfere at the photographic plate with the plane waves arriving from the left. In this case only the upper portion of the circular interference pattern is

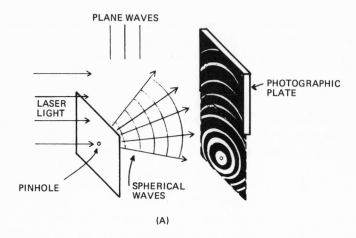

(A)

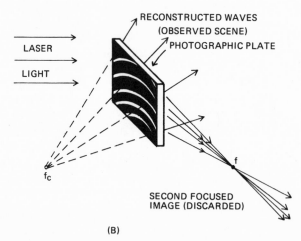

(B)

FIGURE 4.3. The zone-plate equivalence of a hologram is confirmed by a simple hologram of a pinhole. Plane waves and spherical waves diffracted by the pinhole interfere on the photographic plate. The result is a zone plate. In reconstruction, laser light is focused by the zone plate to re-create real and virtual images of the pinhole. Holograms of more complicated images are merely superimposed, independently acting, zone plates.

photographically recorded. When the photographic plate is developed and fixed and then placed in the path of the original pinhole light source is formed at the *conjugate* focal point, f_c. A viewer at the upper right thus imagines he sees the original light from the pinhole. The real image (the focused image) appears at the true focal point as shown; in the usual viewing of a hologram, this second wave set is not used. In this figure, the straight-through, un-deviated waves are not shown.

In Figure 4.4 a similar photographic recording procedure is shown except that in this case the original scene is one having not a single pinhole but three pinhole sources of light, each in a different vertical location and each at a different axial distance from the plane of the photographic plate. We see that each of the three light sources generates its own circular, many-ring

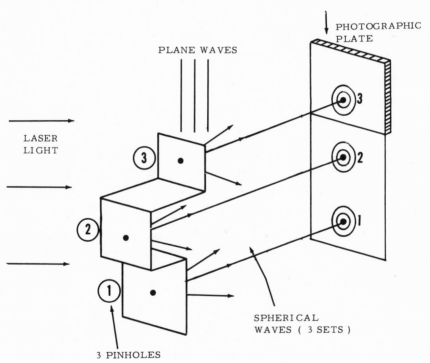

FIGURE 4.4. If the single pinhole of Figure 4.3 is replaced by three separated pinholes, the zone-plate patterns of all three are photographically recorded. When this photograph is reilluminated, all three pinholes are seen in their correct three-dimensional positions. A more complicated three-dimensional scene can be con-sidered as many point sources of light, each generating, on the hologram plate, its own zone plate; each of these zone plates will then reconstruct its source in its original three-dimensional position.

pattern, comparable to the single pattern of Figure 4.3 (in Figure 4.4 only the first two, central, circular sections of these three patterns are indicated). The upper portions of the three sets of circular striations (those encompassed by the photographic plate) are recorded as three superimposed sections of zone plates. When this film is developed, fixed, and then reilluminated, as was done for the single-pinhole recording of Figure 4.3, three sets of upward waves and three sets of focused waves are generated. Of particular importance, from the standpoint of holography, is that virtual images of each of the three pinholes are generated (by the upward, diverging waves). These virtual images cause a viewer at the top right to imagine that he sees three *actual* point sources of light, all fixed in position and each positioned at a different (three-dimensional) location in space. From a particular viewing angle, source three might hide source two. However, if the viewer moves his head sideways or up and down he can see around source three and verify that source two does exist.

4.3. The Hologram of a Scene

All points of any scene that we perceive are emitting or reflecting light to a certain degree. Similarly, all points of a scene illuminated with laser light are reflecting light. Each point will have a different degree of brightness, yet, each reflecting point *is* a point source of laser light. If a laser reference wave is also present, each such source can form, on a photographic plate, its own circular interference pattern in conjunction with the reference wave. The superposition of all these circular patterns will form a very complicated interference pattern, but it will be recorded as a hologram on the photographic plate, as was shown in Figure 4.1. When this complicated photographic pattern is developed, fixed, and reilluminated, reconstruction will occur and light will be diffracted by the hologram, causing all the original light sources to appear in their original, relative locations, thereby providing a fully realistic three-dimensional illusion of the original scene.

The hologram plate itself resembles a window, with the imaged scene appearing behind it in full depth. The viewer has available to him many views of the scene, and to see around an object in the foreground, he simply raises his head or moves it to the left or right. This is in contrast to the older, two-photograph stereopictures which provide an excellent three-dimensional view of the scene, but only one view. Figure 4.5 shows three photographs of one (laser-illuminated) hologram; they are three of the many views a viewer would see if he moved his head from right to left while observing the hologram. For these three pictures, the camera taking them was similarly moved

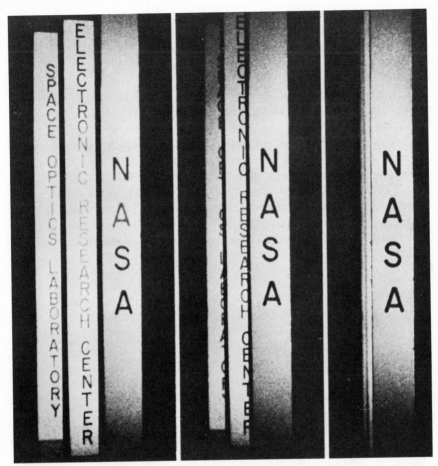

FIGURE 4.5. Three photographs of a hologram being illuminated with laser light. The hologram recorded a scene comprising the three vertical bars of Figure 2.20; for these photographs the camera was moved successively farther to the right, finally causing the rear bars to be hidden by the front bars.

from right to left, fully exposing, in the left-hand photograph, the original three bars positioned one behind the other.

4.4. Parallax in Holograms

In viewing a hologram, the observer is usually encouraged to move his head sideways or up and down so that he may grasp its full realism by observing an effect called *parallax*. In some real scene, more distant objects appear to move with the viewer, whereas closer objects do not. Such effects are very noticeable to a person riding in a train; the nearby telephone poles move past rapidly, but the distant mountains appear to move forward with the traveler. Similarly, the parallax property of holograms constitutes one of their most realistic aspects.

Figures 4.6 and 4.7 are also two different aspects of a single hologram, demonstrating how the parallax property permits the viewer to see behind objects, in this case chessmen.

Because hologram viewers invariably do move their heads to experience this parallax effect, hologram designers often include cut-glass objects in the scene to be photographed. In the real situation, glints of light are reflected from the cut glass, and these glints appear and disappear as the viewer moves his head. This effect also occurs for the hologram and further heightens the realism, as shown in the hologram in Figure 4.8. This is a photograph of a hologram comprising objects which show marked changes in the light re-

FIGURE 4.6. One view of a hologram of a group of chessmen (courtesy of NASA and E. N. Leith).

FIGURE 4.7. Another view of the hologram of Figure 4.6.

flected from various areas as the viewer moves his head; the hologram itself, when illuminated, similarly manifests these light variations. The silver chalice used in this hologram was acquired by the author in 1936 in Bangalore, India, during postdoctoral study with Sir C. V. Raman; also visiting Raman in 1936 was Max Born, recipient of the Nobel prize in physics in 1954 and author of the ageless classic, *Optik*.

4.5. Single-Wavelength Nature of Holograms

One of the basic properties of ordinary holograms (and also zone plates) is their single-frequency nature. Because the design of a zone plate is postulated on one particular wavelength, only waves of that wavelength will be properly focused. Inasmuch as holograms are a form of zone plate, they too suffer from this problem; only single-frequency light waves can properly reconstruct their recorded images. If light comprising many colors is used in

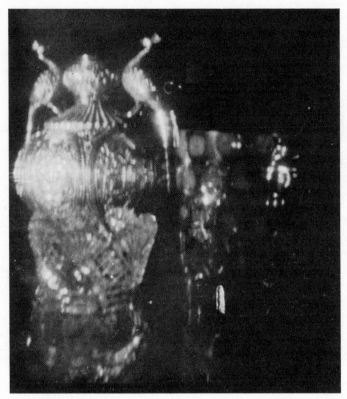

FIGURE 4.8. A photograph of a hologram consisting of the top of a small cut-glass toothpick holder in left foreground, an extensively carved silver object in the left background, and an Early American "thumb-print" glass at right. As the viewer moves his head, all three objects show marked changes in light reflected from various areas.

the reconstruction process, the various colors are diffracted in different directions, and the picture becomes badly blurred. Three examples of this are seen in Figures 4.9–4.11.

4.6. Nonoptical Holograms

Because the basic feature of holography is that of recording a wave interference pattern, holograms can be made using coherent waves of almost any kind. Thus, two sets of radio waves or two sets of sound waves, can be made to interfere, and if records are made of such patterns, they will be radio (or microwave) holograms or acoustic holograms respectively. What were probably the first such holograms to be made were generated acoustically as a means for presenting wave patterns, such as those of Figures 1.6–1.8, 1.22, and 2.7.

4.7. Microwave Holograms

We have seen that a reference wave is needed to form a hologram. Such a reference for generating a microwave interference pattern can be provided

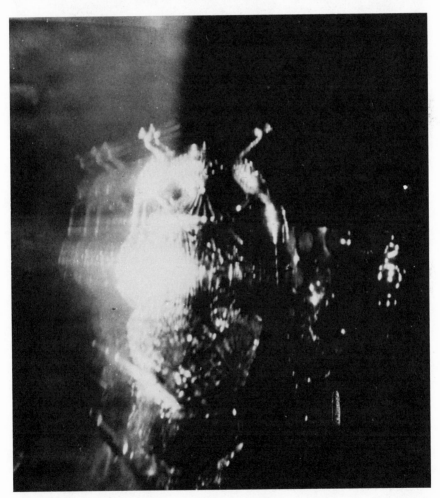

FIGURE 4.9. When the hologram of Figure 4.8 is illuminated with the full light from a mercury arc lamp, serious blurring results.

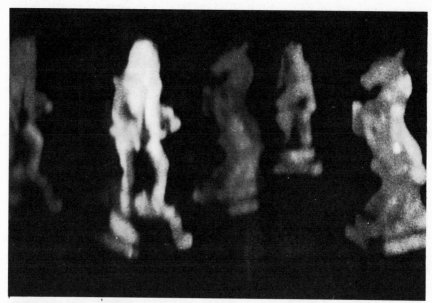

FIGURE 4.10. Two colors of the mercury arc generate two images of each chessman when illuminating the hologram of Figure 4.6. An extra image of a chessman is also shown.

as was shown in Figure 2.6. There the microwaves passing through the lens at the left are the waves of interest, and the coherent waves from the reference feed horn at the right interfere with them at the scanning plane. A camera recorded this interference pattern and the result was shown in Figure 2.5. The lines seen are interference lines, or microwave "fringes," and this record can thus be classed as a microwave hologram.

4.8. Acoustic Holograms

The telephone-receiver pattern of Figure 1.6 is also an interference, or "fringe," pattern (i.e., a hologram), in this case an acoustic one. Because sound waves can be transformed so easily into varying electrical currents (and *vice versa*), the hologram reference wave was injected electronically by the signal coming directly from the electrical oscillator, as was shown in Figure 1.5. The electronically injected reference wave is identical in action to that of a set of plane reference waves and is often used in making acoustic holograms.

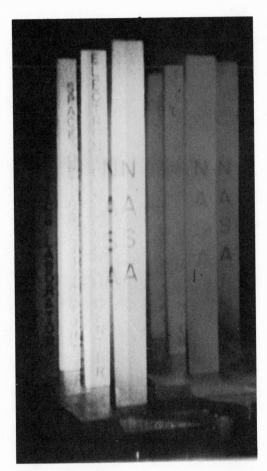

FIGURE 4.11. When the hologram used in Figure 4.5 is illuminated with light from a mercury arc, numerous images appear, corresponding to the various colors (spectral lines) of the arc light.

4.9. Microwave Holograms and Liquid Crystals

Microwave interference patterns can also be portrayed using the recently developed "liquid crystal" technique, in which crystals are employed whose colors are determined by the varying temperature effects introduced by the variations in the strength of the microwave field.[1,2] Figure 4.12 shows such a liquid crystal microwave interference pattern generated by two interfering trains of coherent microwaves moving in opposite directions. As discussed in connection with Figure 3.23, this causes a standing-wave pattern to be generated, and a gratinglike pattern results. We shall discuss liquid crystal holograms at greater length in Chapter 9.

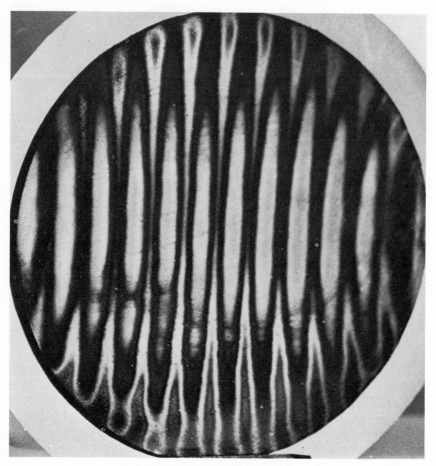

FIGURE 4.12. The standing waves illustrated in Figure 3.23 can be portrayed using oppositely traveling microwaves and a new development called liquid crystals.

4.10. Liquid-Surface Acoustic Holograms

We have discussed the procedure for making two-dimensional acoustic holograms *photographically* by recording the sound-wave interference pattern, as in Figure 1.6. For holograms of actual objects, the sound interference pattern must also be recorded by transforming the acoustic pattern into a light-wave pattern. The photograph must then be reduced in size and viewed with laser light for optical reconstruction of the scene.

Dr. Rolf Mueller of the Bendix Research Laboratories pioneered in using, as an ultrasonic hologram surface, a liquid–air interface (Figure 4.13). A coherent reference wave is directed at this liquid surface,[3] and it becomes

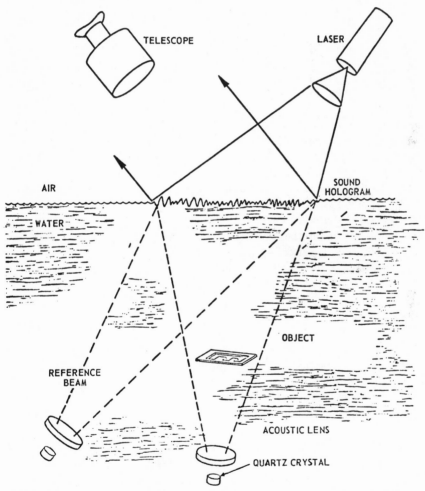

FIGURE 4.13. When the waves of interest and the reference waves of an underwater acoustic hologram interfere at the water surface, the ripples constitute the hologram, and immediate reconstruction can be brought about by shining laser light on these interference ripples.

the "recording" surface for the hologram interference pattern. Because the surface of a liquid is a pressure-release surface, i.e., a surface which rises at points where higher-than-average sound pressures exist, the acoustic interference pattern transforms the otherwise plane liquid surface into a surface having extremely minute, stationary ripples on it. When this rippled surface is illuminated with coherent (laser) light, an image of the submerged object is reconstructed. We shall discuss this subject further in later chapters.

4.11. Hologram Equations

A hologram, whether it is an optical, microwave, or acoustic one, is a recording of an interference pattern formed by superimposing a reference wave upon an object-scattered wave. Because the interference depends upon the relative amplitudes and phases of the two waves, highly coherent (single-frequency) waves must be used in order to keep the phase relationship constant during the generation of the photographic record of the hologram. In making holograms in which scanning procedures are used, the time taken to record the hologram may be quite long. Wave coherence must be maintained over the entire period if a satisfactory hologram is to be formed.

If we call E_1 the reference wave (assumed, for simplicity, to be plane), and E_2 the object-scattered wave, a sampling of the acoustic interference pattern located in an xy plane will yield an intensity variation

$$
\begin{aligned}
I(x, y) &= (E_1 + E_2)(E_1 + E_2)^* \\
&= |E_1|^2 + |E_2|^2 + E_1 E_2^* + E_1^* E_2
\end{aligned}
\tag{4.1}
$$

where the asterisk indicates the complex conjugate. In converting this intensity variation to a photographic record, the last two terms, $E_1 E_2^*$ and $E_1^* E_2$, are the only contributors, i.e., they are the relevant signal-bearing terms. On reconstruction, the first two constitute the zero-order diffracted terms. When the photographic record is illuminated with a third (plane) wave E_3, it will become, in passing through the record, a modulated wave E_4, specified by

$$
E_4 = E_3 E_1 E_2^* + E_3 E_1^* E_2
\tag{4.2}
$$

Both E_3 and E_1 are plane waves, so that $E_1 = E_1^*$, and both product terms $E_3 E_1$ and $E_3 E_1^*$ are constants. Equation (4.2) thus states that the object-scattered wave E_2 and its conjugate E_2^* are reconstructed. This is the same as saying that virtual and real images of the object are reconstructed. If non-identical reference and reconstructing waves are used, a magnification factor enters.

4.12. References

1. C. F. Augustine and W. E. Kock, Microwave holograms using liquid crystal displays, *Proc. IEEE* **57** (3), 354–355 (1969).
2. C. F. Augustine, C. Deutsch, D. Fritzler, E. Marom, Microwave holography using liquid crystal area detectors, *Proc. IEEE* **57** (7), 1333–1334 (1969).
3. R. K. Mueller and N. K. Sheridon, Sound holograms and optical reconstruction, *Appl. Phys. Lett* **9**, 328 (1966).

5

PROPERTIES OF HOLOGRAMS

Some features of holograms, when they were first described in the literature, were considered so surprising that numerous expressions of disbelief were voiced publicly; a few examples of such questioning will be included in the following discussion of hologram properties. Probably the most spectacular characteristic of holograms, and the one undoubtedly responsible for the tremendous popular interest they have evoked, is their ability to create an extremely realistic illusion of three dimensions.

5.1. Three-Dimensional Realism

The three-dimensional property of holograms is difficult to demonstrate, lacking an actual hologram with its needed illumination. We have noted that some idea of this effect can be gotten by looking at several different photographs of the same, properly illuminated hologram, such as in Figures 4.5–4.7. Today, fairly simple and fairly inexpensive equipment is available for viewing holograms (even in the home) and, thereby, for observing not only the surprising three-dimensional effects, but also other interesting properties such as those present when a lens is included as one of the objects in the hologram (we shall discuss such lens holograms shortly). The supplier (Holex Corp.) of the viewing equipment offers a variety of interesting (and inexpensive) holograms, including one which portrays a watch located behind a lens. In the (Holex) hologram reconstructed in Figure 5.1, a row of dominoes is seen; movement of the viewer to the left causes the rear dominoes to be hidden by the front one.

We noted in connection with Figure 4.3 that both a virtual and a real image are generated by the zone-plate hologram of a point source, and that the *real* image is usually discarded (not used). We shall see that the reason for discarding this image is because it had the peculiar property of being pseudoscopic, a property which, however, can be corrected (as shall be described later). This correction was done for one of the most spectacular (and notorious) holograms of recent times; and for it, the *real* image was used. As was noted in discussing Figure 4.3, the real image exists on the *viewer's* side of the hologram, not on the opposite side as in the case of the image in Figure 4.2, where the hologram appears as a window, with the image positioned behind it. The image of the hologram referred to is shown in Figure 5.2. It comprises a woman's outstretched arm, with her hand holding a string of very valuable diamonds. The hologram itself, being *behind* the image, was placed just to the rear of the bulletproof glass window (Figure 5.3) of a famous jewelry store (Cartier's) on Fifth Avenue in New York City, and the image of the woman's arm holding the diamonds extended out over the sidewalk of Fifth Avenue. The effect was so real to the viewer that one woman, according to a report

FIGURE 5.1. A hologram viewer using filtered white light. (Courtesy of Holex Corp.)

FIGURE 5.2. A hand "offering" $100,000 worth of gems—a diamond ring and bracelet—suspended via a holographic system over the sidewalk next to Cartier's on Fifth Avenue in New York.

published in the journal *Science*,[1] attempted to strike the image several times with her umbrella, maintaining that "it was the work of the devil."

In viewing a hologram, the three-dimensional illusion is far more realistic than Figures 4.5–4.7 can convey. The viewer quickly realizes that much more information about the scene is furnished by a hologram than by other three-dimensional photo processes, such as by stereophotography, involving a pair of stereophotos. In the hologram reconstruction, the viewer can inspect the three-dimensional scene not just from one direction, as in stereophotography, but from many directions. Because this property of complete realism has undoubtedly given the development of holography great impetus, let us review the chronology of three dimensionality in hologram development.

As early as 1949 Gabor wrote that "the photography contains the total information required for constructing the object, which can be two-dimensional or three-dimensional." [2] However, Gabor used two-dimensional transparencies for his objects and said little as to how the recording and

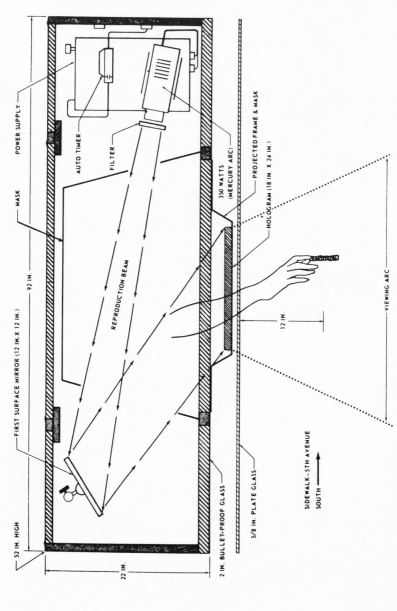

FIGURE 5.3. The arrangement of the holographic system used to reconstruct an image through Cartier's 2-in.-thick bullet-proof display window. Holoconcepts of America took the existing vaultlike Cartier window enclosure and utilized it for the holographic display shown in Figure 5.2. The hologram was shown with a reference beam angle that accommodated the inner perimeter of the display area.

reconstruction of three-dimensional objects would be accomplished. Accordingly, when, in December, 1963, Leith, who was an esteemed contributor in the radar field, first used lasers in holography, he likewise employed two-dimensional transparencies for his objects.[3]

E. N. Leith thus pioneered the first laser holograms and the first offset holograms, but his first objects and images were still two-dimensional. Also, he discussed his early hologram experiments in terms of the optical processing technology of coherent (synthetic aperture) radar; we shall note later the close relationship between holography and this form of radar. Later, Leith and his group contributed extensively to three-dimensional holography. Figure 5.4, taken at the NASA Electronics Research Center when the author was its Director, is a photograph of Leith and a very spectacular hologram of his being reconstructed by the laser beam at the left.

George Stroke (Figure 5.5), a recognized authority on the ruling of precision optical gratings, commenced quite early in looking upon holograms (as this book has) as diffraction devices. In a May 1964, set of lecture notes at the University of Michigan,[4] he described how light waves, reflected onto a photographic plate by two adjacent, slightly tilted plane mirrors, generate a photographic grating (Figure 5.6), and how, when a three-dimensional object is substituted for one of the two mirrors, a three-dimensional hologram results. It was not easy for some to accept this generalization. Thus, in the discussion period following a lecture by Stroke on holography in Rome in

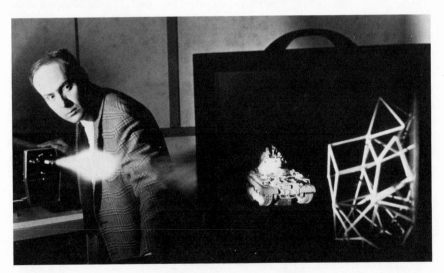

FIGURE 5.4. Emmett Leith, the first to use lasers in holography, illuminates one of his holograms at NASA's Electronics Research Center (courtesy of NASA).

FIGURE 5.5. The author (left) and George Stroke, a pioneer in making laser holograms, and who first proposed the term "holography" (taken at the 1973 NATO Advanced Study Seminar in Capri).

September 1964, an eminent Italian scientist remonstrated: "the light beam cannot carry information about a three-dimensional object because this is described by three degrees of freedom, whereas a light beam has only two degrees of freedom." [5] Although the logic of this objection appears at first sight quite reasonable, we can assure ourselves by, for example, a consideration of Figure 4.4, that three-dimensional information *can* be recorded on a two-dimensional surface by means of the hologram (zone-plate) process. We shall see that a similarly surprising situation can exist in certain forms of holographic synthetic-aperture radar, where there, a *one*-dimensional record can provide full information about a two-dimensional *plane*.

The three-dimensional realism of certain particular holograms is often difficult to understand. Thus, Figure 5.7 shows a hologram recording a scene which includes a small, very bright reflecting point having an opaque screen in front of it. In the actual scene, a viewer positioning himself so as to be below

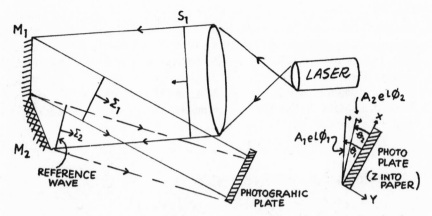

FIGURE 5.6. A sketch from George Stroke's 1964 Lecture Notes, showing how two plane waves can generate a photographic grating, permitting the reference wave to *reconstruct* the other plane wave.

the shadow demarcation line X–X would have his view of this point source quite understandably blocked by the opaque screen. In the reconstruction, however, the skeptic would say that the screen is actually not there; there is only an *image* of the screen, a ghostlike, evanescent figment of the imagination. So how can this *image* of a screen block the reconstructed light issuing

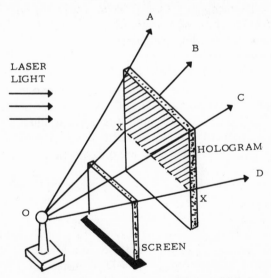

FIGURE 5.7. A hologram of a point source of light which is partially hidden from the hologram by an opaque screen presents a paradox: On reconstruction, can the ethereal, nonexistent screen still hide the imaginary light source?

from the point? Here again, the zone-plate analogy provides the answer. In the recording process, only a portion of the reflection point's zone plate is recorded on the photographic plate (it is shown in Figure 5.7 as a shaded area). The remainder (below line $X–X$) having been blocked by the screen, is missing. Accordingly, as far as the luminous point itself is concerned, its hologram "window" ends at the line $X–X$, and its (partial) zone plate can only diffract laser light in directions encompassed by the pyramid $OABCD$. The point light source *can* be blocked by an apparition!

5.2. Holograms and Photographs

There are two ways in which the hologram photograph differs from an ordinary photograph. When the usual photograph is taken of a scene, a negative is first made, then the dark and light areas of this negative are reversed to form a positive print which portrays the scene in its original form. In holography either the positive or the negative version of the hologram generates the identical three-dimensional illusion. This property follows from the similarity between holograms and zone plates; we noted that zone plates and gratings perform equally well if their dark (blocking) areas and their bright (transparent) areas are interchanged (Figure 3.9 and 3.11). This process is equivalent to changing a positive print to a negative one or *vice-versa*; it constitutes a major change in photography but no change at all in holography.

The second way in which holograms differ from photographs is in the appearance of the photographic plate. When a hologram is held up to the light, hardly any pattern is seen, certainly none which is indicative of the scene recorded, as in a photographic negative. The hologram appears to be a fairly uniform gray sheet (Figure 5.8), and it reveals none of the characteristics or features of the scene recorded until it is properly illuminated. The interference pattern recorded involves the extremely fine fringes of light wavelengths; these are so small that they are only visible when highly magnified.

5.3. Lens Action

One interesting proof of the realism of holograms is provided when a *lens* is included in the recorded scene.[6] Then, instead of a sideways motion of the viewer for observing the parallax effect, a motion toward and away from the hologram gives him the expected enlarging and contracting of objects behind the lens. This phenomenon is shown in Figure 5.9, in which two

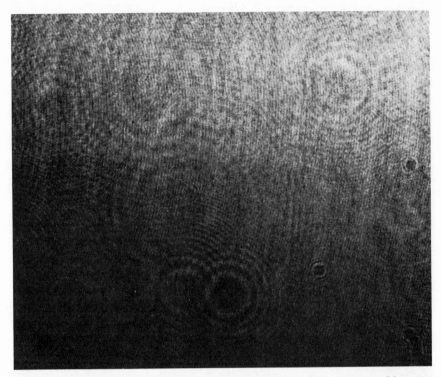

FIGURE 5.8. When a hologram is viewed in ordinary light, no resemblance to the actual recorded scene is evident.

different photographs of the same hologram are portrayed. The scene recorded contains a magnifying glass placed in front of some lettering. The top photograph was taken with the camera close to the hologram; for this position the magnifying glass includes the letters *E* and *S* and a portion of the upper letter *R*. In the lower photograph the camera was moved some distance back, and a change in magnification is seen. This magnification effect is heightened with increased distance

5.4. A Stereohologram

A particularly interesting hologram involving lenses was made at the Bendix Corporation's Research Laboratories and exhibited in March 1966 at the Convention of the American Congress of Surveying and Mapping and the American Society of Photogrammetry in Washington, D.C. A side view of the

FIGURE 5.9. When a lens is included in the recorded scene, objects behind the lens seem larger when, as in the bottom photograph, the camera is farther away from the letters.

scene involved is shown in Figure 5.10. The objects consisted of a pair of stereoviewing lenses positioned so as to permit a viewer, placing his eyes in close proximity to the stereolenses, to observe a three-dimensional (stereo) view of a model of some mountainous terrain. The hologram was made with the hologram plate in very close proximity to them. To view the reconstructed scene, the observer positions his eyes as though the two viewing lenses are actually in place. When he thus looks directly into the *imaginary* lenses he imagines he sees a magnified view of the terrain in three-dimensional stereo, just as he would through the original stereoviewing lenses. One photograph

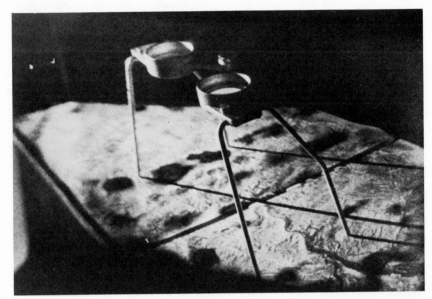

FIGURE 5.10. These stereo-viewing lenses are positioned to permit a viewer to observe a three-dimensional (stereo) view of the model of mountainous terrain below it.

FIGURE 5.11. One photograph of a reconstructed hologram of the scene depicted in Figure 5.10.

FIGURE 5.12. A closeup of the view from one lens of the hologram of Figure 5.11.

of the illuminated *hologram*, showing the two lens-imaged views, is shown in Figure 5.11, and a second, taken with the camera close enough to match the position of the viewer's eyes, is shown in Figure 5.12; the greater detail of the viewed terrain is evident in Figure 5.12.

5.5. Focused-Image Holography

Another interesting use of lenses in holograms was made in 1966 by NASA scientist Lowell Rosen in the process now called focused-image holography.[7] The procedure is shown in Figure 5.13; a lens forms a real image of a group of objects located behind the lens, and a hologram is then made of this (upside-down) real image. The photographic plate can even *straddle* objects separated longitudinally in the real lens *image*, i.e., the image of one object can be positioned behind the photographic plate and the image of a closer object positioned in front of the hologram plate. When this hologram is developed and properly illuminated, the viewer sees one object behind the hologram plane and another standing out in space in front of the hologram.

When the focused-image-hologram process was first described, questioning was again evident. It was assumed by some that the presence of a lens would automatically cause a collapsing of the image onto the photographic plate. A joint publication on this subject by Kock, Stroke, and Rosen [8] noted that it had generally "been accepted without question in the past that the 'third dimension' in a conventionally-focused photographic system would be irretrievably lost in the recording on simple, two-dimensional photographic plates or film." It continued: "In the normal photographic process, the camera lens forms real images of the objects of a three-dimensional scene, and these images exist in three-dimensional space; however, the ordinary photographic film causes these images to be collapsed into a single plane." We note that if the plate in Figure 5.13 had simply been exposed with the light from the objects (i.e., as in a true photograph, without the simultaneous use of the reference beam, *B*), the result would have been an ordinary photograph, and the images *would* have been collapsed in the resulting picture. The hologram process, however, retains the longitudinal, three-dimensional image separation.

It is of interest to consider what would happen if, as shown in Figure 5.13, an opaque sheet were to be positioned, so that it, too, straddles the two real images. In the original situation, this sheet, being opaque, would prevent the lens from generating the actual image of the pyramid; also, a viewer who

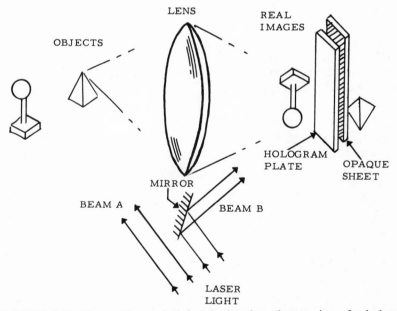

FIGURE 5.13. Focused-image holography involves the creation of a hologram of upside-down images projected by a lens. When this holographic plate is developed and illuminated, the viewer sees one object behind the hologram and another standing out in space in front of it.

could see both images if this sheet were removed, would see neither. The hologram photographic plate, however, records the wave pattern existing at its plane, and this original wave pattern possesses the intrinsic ability to form the pyramid image (as it would if the opaque sheet were removed). Accordingly, the presence of the opaque sheet has no effect on the hologram recording, nor on its later ability to establish real images of both objects in the reconstruction process. The hologram has recorded an image, which, with the opaque screen in place, does not exist, and it can later establish this "nonexistent" image in its reconstruction process (with the screen then removed, of course).

Interesting effects are obtained in focused-image holograms when an actual object is placed adjacent to the upside-down real images. For it to be recorded, it must, of course, be *behind* the hologram plate, a restriction which, we just saw, does not apply to the real images. The parallax effects are not the same for this true object and the imaged objects, and a most intriguing and confusing scene results when the hologram of such a combination is viewed. At about the time that Rosen was first experimenting with focused-image effects, Gabor visited the NASA center in Cambridge; a photograph was taken there of Rosen, Gabor, and the author.[9] The frontispiece is a more recent photograph of Nobel Laureate Gabor, taken at the Third U.S.–Japan Holography Symposium held in January 1973 at which George Stroke and the author were U.S. cochairmen.

5.6. Reconstruction with a Small Portion of a Hologram

We observed in connection with Figure 3.18 that focusing devices such as paraboloidal surfaces or zone plates can achieve a focusing effect even if only a portion if their focusing surface is utilized. Because a hologram is a form of zone plate, it too can achieve certain of its effects if only a small portion of the original hologram is used in the reconstruction process. Figure 5.15 portrays this property for a hologram zone plate; both the full area $ABCD$ and its smaller portion $A'B'C'D'$ generate real and virtual images.

Let us consider what effect this size reduction has on the zone plate of Figure 5.15. Because less light falls on the smaller area $A'B'C'D'$, less light is diffracted toward the focal point f, and the focused, real image is, therefore, not as bright. In a similar way, the real image of a full *hologram* is not as bright, the brightness reduction being in proportion to the area. The original scene is still imaged however, and it can be portrayed on a white card located at the focal plane.

The size reduction also lowers the light intensity for the virtual image of a hologram, but in addition, the realism of the image is reduced. Thus, in

FIGURE 5.14. NASA's Lowell Rosen (right) shows holography inventor Dennis Gabor (center) a reconstructed hologram while the author looks on (courtesy of NASA).

Figure 5.15, an observer viewing the virtual image of the point source of light will be required, for the smaller area $A'B'C'D'$, to position himself more carefully in order to see the source through this new, smaller window, and this smaller window reduces appreciably the realism he would obtain in a full hologram through the parallax effect. Nevertheless, with the smaller window, the viewer can still move his head about and see all of the objects in the scene, the *window area* he is using does not move with him, and he merely sees small, individual, almost two-dimensional views. In the case of a glass window and a real scene, the viewer's ability to see around objects in the foreground is

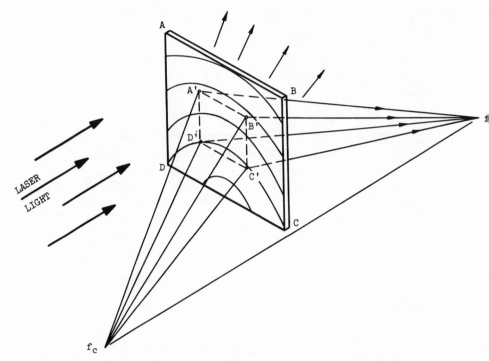

FIGURE 5.15. Only part of a hologram is necessary to reconstruct a scene in its entirety.

also obviously limited when the window size is reduced to a significantly smaller area. We shall note that in the case of nonoptical holograms (such as microwave or acoustic holograms) the much longer wavelengths involved severely restrict the parallax and three-dimensional effects. The long wavelengths are responsible for a much smaller viewing window.

When a complete optical hologram is available but is illuminated only by a small-area "spot beam" of laser light, other effects are produced. As this small-area beam of light is moved about over the hologram area, the viewer of the virtual image can only see all views (from all angles) of the recorded scene by moving his head about. A white card placed at the focal area will similarly display the *real* image on it of a series of changing (two-dimensional) views of the scene. This latter ability to change, for an observer, the direction from which he views the original scene could have interesting applications. An aircraft pilot, attempting to land his plane in fog could have presented to him a prerecorded hologram picture of the airport runway from the very angle (up, down, or sideways) that he would be observing it if the weather were clear.[11]

5.7. Pseudoscopy in the Real Image of a Hologram

In the discussions of hologram zone plates, it was noted that they can function as converging lenses, so that the real image they form can be recorded on film just as the real image of a camera lens is. A hologram of a scene also produces a real image of the scene, and this image can be viewed on a white card placed at the focal area, just as the real image of a camera lens can be focused on a ground-glass screen. When the scene recorded by the hologram is a two-dimensional transparency, the real image it reconstructs is also two dimensional, and it is then indistinguishable from the real image that a lens would form of that transparency. When, however, the original scene is three dimensional, the real image of a hologram is also three dimensional; however, it differs quite significantly from the real image which a lens forms of the same three-dimensional scene.

This effect is shown in Figure 5.16; the viewer is assumed to be at the right of the figure.[12] As an object is moved further to the left of the focal point of the lens, as from B to A, the position of its image likewise moves to the left, as from B' to A'. Objects A' and B' in this lens-produced real image thus bear the same distance-from-viewer relationship to the viewer as the actual objects A and B do.

For the hologram real image this is not the case. In the recording process, luminous object A forms its own zone plate, and the real (and virtual) focal point of this, when developed and illuminated, is therefore at the equidistant point A'. Similarly, object B forms its own zone plate, with the real focal point at *its* equidistant point B'. For the viewer of this real image, objects originally in the rear appear to be in the foreground and *vice-versa*. This reversal of forward and back objects has been given the name *pseudoscopy*; the image is said to be pseudoscopic and because such images are rather confusing, little attempt had been made to utilize the *real* image of hologram for viewing purposes. To overcome this effect, scientists at KMS Industries in Ann Arbor* used a circularly symmetrical object (a champagne glass) as a hologram subject. In the reconstruction process they presented the viewer with the real image, which stood out in front of the hologram as sketched in Figure 5.17. Because of the symmetry of the champagne glass, the exchange of front for rear was not significant, and the viewer imagined he was seeing a normal, nonpseudoscopic image, standing out *in front* of the hologram.

Others proceeded along different paths to obtain " out-in-space " images. We have already mentioned one process; that of focused-image holography.

* This hologram, made by N. Massey and E. Champagne of KMS Industries was exhibited at the American Optical Society Section Meeting, January 27, 1968, in Ann Arbor, Michigan.

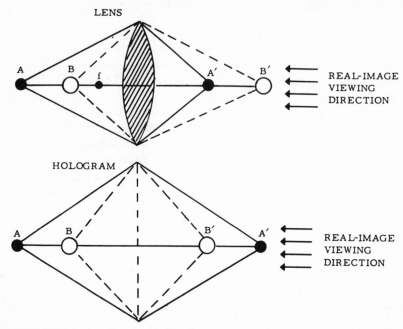

FIGURE 5.16. Pseudoscopy is illustrated in this diagram. Objects imaged by a lens retain their original relative positions. When a hologram is made, each object forms it own zone plate and, on reconstruction, the real and virtual images are positioned equidistant from the hologram; pseudoscopy or image inversion thereby occurs.

Another procedure used was to make a *second* hologram of the real (pseudoscopic) image of a first hologram.[3] The real image of this second hologram then became a "pseudoscopic-pseudoscopic" image (a doubly reversed image), in other words, a normal image. In this process any scene, including unsymmetrical ones, can be used as a subject. It was this method which was used to generate the image of a woman's arm, discussed earlier, stretched over the sidewalk of New York's Fifth Avenue (Figure 5.2).

5.8. Image Inversion

As suggested in Figure 5.17, both the real and virtual images can be presented to the viewer. Usually, as in Figure 5.17, they are at widely different angles of view, and the viewer cannot see both simultaneously. He must either reposition himself to see the second image, or he must reposition the hologram.

Figure 5.18 shows how a simple rotation of a hologram can often accomplish this. When this (point source) hologram is rotated a half turn as shown, a viewer placed so that he originally saw one of the images now would see the other.[14] A hologram made in this way of an actual scene would also behave in this manner. Because the viewer is usually at some distance behind focal point f, the real image, having passed through the focal point, is inverted, as for any focusing lens. Were he to have changed his own position so as to see the other image, he would have observed that the two images are inverted relative to one another. In switching his view from one to the other by a rotation of the hologram, however, the second image is rotated from top to bottom. Thus, if the image he viewed first was rightside up, the new image would also be rightside up, even if he turns the hologram upside down. Rotations about other axes also cause curious effects.

When in 1965, the author published on this phenomenon,[14] a wide response again resulted. A former colleague sent a letter saying, "Beware! Other people's holograms don't work that way," an M.I.T. professor explained the phenomenon very elegantly with Fourier transforms, and a third correspondent wrote "the . . . properties . . . are exactly what one should expect." [15]

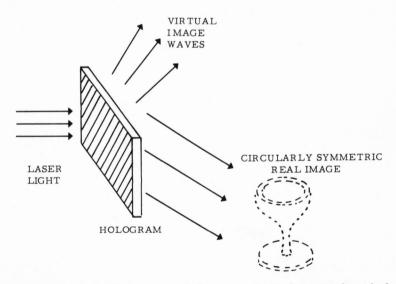

FIGURE 5.17. A circularly symmetric image can be made to stand out in front of the hologram very realistically despite the pseudoscopy associated with the real image of a hologram.

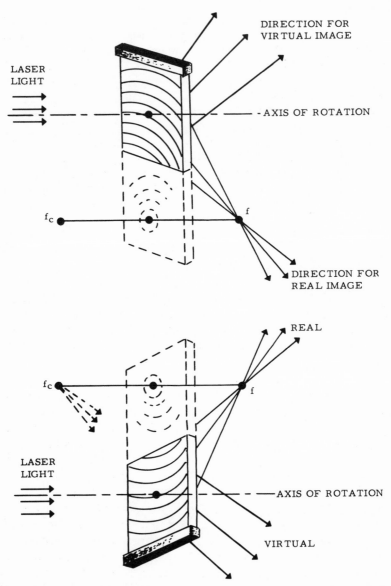

FIGURE 5.18. By rotating certain holograms, the viewer can see real and virtual images alternately.

5.9. Color Holograms

One of the properties of ordinary holograms (and zone plates) which limits their usefulness is their single-frequency nature. We saw earlier that the design of a zone plate is postulated on one particular wavelength, and that for a given zone plate only waves of one wavelength will be properly focused. Inasmuch as holograms are a form of zone plate, they, too, suffer from this problem; only single-frequency light waves can properly reconstruct their recorded images. If light comprising many colors is used in the reconstruction process, the various colors are diffracted in different directions, and the picture becomes badly blurred.

We noted in connection with Figure 4.9–4.11 that when holograms made using single-frequency laser light are illuminated with light from a mercury arc lamp, the holograms cause each color to be diffracted in a different direction, so that objectionable blurring results.

This effect in holograms, whereby different colors are diffracted different-ly, makes it difficult to form a three-color hologram simply by exposing the photographic plate to three different-colored laser illuminations. In recon-struction, the illumination by *three* laser beams of this "compound" holo-gram would, in general, create *nine* images, only three of which would be in proper registry. A special technique which is employed in making *reflection* holograms is one way to overcome this multi-image difficulty. The pro-cedure is that described earlier for making reflection zone plates (Figure 3.23–3.25).

In this process, the *thickness* of the photographic emulsion is utilized to produce a *longitudinal* recorded interference pattern. As we noted this concept was first demonstrated by Lippmann.[16,17] Lippman first produced extremely fine-grain photographic emulsions, with silver bromide grains smaller than a wavelength of light. He then put such a plate in a camera, with the emulsion facing the back of the camera, and backed up the emulsion with mercury as a mirror. The effect was most spectacular. The light waves falling in from the front side interfered with the waves reflected from the back and, as also occurred in Figure 3.23, formed a set of standing waves, parallel to the mirror, with maxima half a wavelength apart. Very fine silver grains were precipitated in the maxima. Each of these "Lippmann layers" scattered light a little, but their effect reinforced one another only when all these wavelets were in phase, and this occurred only for the original color, i.e., that light having the *wave-length* corresponding to twice the layer spacing. The Lippmann plates selected from *white* light the original color, and reflected only this to an appreciable extent. Even more surprising, they reproduced faithfully a *mixture* of pure colors because each color produced its own set of Lippmann layers, and these operated additively, without disturbing one another. In 1962, Van Heerden drew attention to the immense storage capacity of deep emulsions.

In 1962, the Soviet scientist Y. N. Denisyuk conceived of the idea of combining Gabor's holography with Lippmann's color photography.[18] Denisyuk's idea was to combine the Lippmann process with holography by substituting for the reflected wave a reference wave, falling in from the back of the emulsion. The object wave and the reference wave again form standing waves, but unlike in Lippmann's camera, they are not parallel to the emulsion surface, but form equal angles with the two waves. They produce, as it were, a succession of partial mirrors, which, on illumination by the reference beam alone, reflect in the direction of the original object ray. In other words a "deep," or "volume," hologram is formed which shows the object in its original spatial position. Moreover, the second image, which was always a disturbance in transmission holograms, vanishes altogether in these "reflection holograms."

Denisyuk had no laser at his disposal; also, his illumination of the object *through* the plate presented some serious limitations comparable to the original in-line arrangement of transmission holography. Later, Stroke and Labeyrie[19] improved the process by illuminating the objects *directly* while the reference beam remained incident on the back side of the plate. They showed, moreover, that these holograms could be made to reconstruct natural colors when viewed in white light.

5.10. Requirements on Film Properties

Light wavelengths are extremely short, and the recording, on film, of a light-wave *interference* pattern demands film which has a very high resolution capability. All photographic films have a coating or emulsion placed on a glass or flexible-film base. The emulsion is a mixture which usually includes a form of silver salt as its light-sensitive ingredient. This mixture of materials can be finely or coarsely divided and the fineness of the grains of the light-sensitive ingredient determines how much detail can be recorded photographically. When a recorded film is enlarged, one with coarse grains shows blobs which limit the amount of detail recognizable. Fine-grain film permits a much greater enlargement before such a limitation of detail becomes evident.

The film emulsion most widely used in early laser-hologram experiments, and one widely used still, is the Kodak 649F spectroscopic emulsion. It and others now available, such as Agfa, are capable of recording up to 1500 lines/mm. In holography, fine-grain emulsions are necessary because most holograms possess a very fine fringe spacing. For the zone-plate design shown in Figure 5.1, the higher zones are at very wide off-axis angles from the focal point; their spacing approaches one wavelength of light. In other holograms, similarly, if the angle between the reference beam and the widest angle of the waves from the scene exceeds 60°, the fringe spacing is very close to one light wavelength. At the wavelength of helium–neon (He–Ne) lasers, this would be approximately 1500 lines/mm. It has been found that the 649F and other presently available emulsions can satisfactorily accommodate this spacing; in fact, these emulsions are even satisfactory for reflection-type (white-light) holograms which have, as we saw for reflection zone plates, exposed areas separated longitudinally within the emulsion which are spaced only a half wavelength apart (3000 lines/mm). The emulsion speed of 649F is only 0.05 (ASA rating) at the red-light region of the Helium–Neon gas laser (6328–Å wavelength). For this emulsion, and for such a laser rated at 0.1–W output, an exposure time of approximately 1 minute is required.

During the exposure time, the interference pattern to be recorded must not be allowed to change (as would occur if even fantastically small relative movements took place between objects of the scene or between them and the reference beam). Accordingly, in making the hologram, the objects are usually placed on something which will not move or vibrate during the exposure time, such as a several-ton block of stone or marble. A shift of even 1/100,000 of an inch of any object will change the relationship of the wave crests and troughs, thereby moving and blurring the interference pattern on the film and ruining the hologram. If the angle between reference beam and the light from the scene is kept small, the fringe spacing becomes much larger; faster film can then be used. Agfa film Agfapan FF can resolve 500 lines/mm and has an ASA rating of 25.

5.11. Information Content

The spectacular property of holograms, which provides the viewer with information about the scene recorded, is also exactly the property which has unfortunately limited the use of hologram principles in many interesting applications. Such applications include three-dimensional movies and three-dimensional television, fields which would obviously benefit if the realism of holograms could be imparted to them. The large information content of a hologram is inherent in the extremely fine detail which must be recorded on the holograph film. As we saw, this detail, of about 1500 lines/mm, is only available in very special photographic film. Present-day television systems employ a far coarser line structure, so that the outlook for using holograms in television is very bleak.

Several methods have been suggested for reducing the information content of a hologram without sacrificing completely some of its interesting properties, but this road is a very difficult one. Television pictures in the United States have approximately 500 lines vertically and 500 dots horizontally. One television picture (corresponding to one frame of a movie film) thus has an information content corresponding to 500 × 500, or 250,000, dots. A 200-mm × 200-mm hologram (approximately 8 in. × 8 in.) having 1500 fringes (lines)/mm, would have the equivalent of 200 × 1500, or 300,000, lines vertically and 300,000 dots horizontally, with a total information content corresponding to 90,000,000,000 dots (90 *billion* dots). The ratio between the information in an 8-in.-square hologram and that in a U.S. television picture is thus 360,000. To reduce the information content of a hologram by a factor that large would truly be a remarkable accomplishment.

Procedures which have been proposed to permit at least some reduction in the hologram information content include one (suggested by the author [20]) which involves retaining many tiny areas (perhaps 300–500 in each vertical line and a similar number in each horizontal line) and discarding the remaining much larger areas (Figure 5.19). Each retained area, though extremely small, would still contain several light fringes, [21,22] and the assemblage would therefore still possess the zone-plate character of the original hologram.

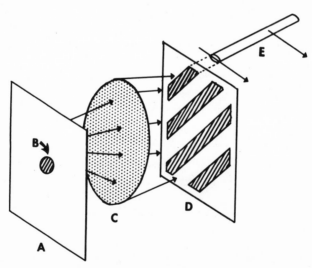

FIGURE 5.19. By dividing a hologram into many individual sections (such as the one at *A*), then enlarging a tiny portion, *B*, of that section for television transmission of the fringes (at *D*) by the electron beam *E* of a vidicon, the transmission bandwidth can be sizably reduced.

Another technique retained one full, but extremely narrow, horizontal strip of a hologram (0.01 of the original height), and then a new hologram was formed by repeating this one strip many times in the vertical direction. The new hologram displayed good horizontal parallax, and, accordingly, for the viewer's two, horizontally-positioned, eyes, provided quite realistic three-dimensional effects. It lost, in the process, the vertical parallax originally observable by a vertical motion of the viewer.

5.12. Holograms and Coherent Radar

The reconstruction of phase information in holography sets it definitely apart from photography and shows that, in spite of the similarities, holography is not just a variant of photography, but it is rather a basically new process. This fact is significant because two new technologies, radar and sonar, have patterned themselves along the lines of the older classical science of photography, with its extensive use of lenses and reflectors. These technologies have accordingly inherited the properties and limitations of photography. A new form of radar, coherent radar,[23] is free from these limitations because it is a form of holography, and the awareness of the basic differences between holography and photography, and similarly between coherent radar and ordinary radar, has led to important developments in radar and sonar.

Extensive contributions to coherent radar were made in the late 1950's at the University of Michigan by a team under the leadership of the American scientist Louis Cutrona,[24] and one of his team members was E. N. Leith. Because the relationship between coherent radar and Gabor's holography is so close, one could almost say that holography was reinvented by Cutrona and his team. They generated zone plates (one-dimensional ones), they employed offset procedures to separate the diffracted components, and they used laser optics to regenerate the radar image. Fortunately for holography, one team member, Leith, ingeniously extended his coherent-radar knowledge to optical holograms, and the first visible-light laser hologram thereby came into existence. We shall discuss this subject at some length in Chapter 9.

5.13. References

1. *Holosphere* **Mar.**, 4, 5 (1973); *Science* **180**, 484–485 (1973).
2. D. Gabor, *Proc. R. Soc. Lond. Ser. A* **196**, 464 (1949).
3. E. N. Leith and J. Upatnieks, *J. Opt. Soc. Am.* **53**, 1377 (1963).

4. G. W. Stroke, An introduction to optics of coherent and non-coherent electromagnetic radiations, Laser summer course at The University of Michigan, Ann Arbor, Mich. (May 1964). This is now largely included in the book, *An Introduction to Coherent Optics and Holography*, Academic Press, New York, 1966.

5. G. W. Stroke, Wavefront reconstruction imaging, invited lecture given at Atti del Convegno sui Campi Magnetica Solari, Florence, Italy, Sept. 14, 1964.

6. W. E. Kock, L. Rosen, and J. Rendiero, *Proc IEEE* **54**, 1985 (1966).

7. L. Rosen, *Proc. IEEE* **55**, 79 (1967).

8. W. E. Kock, G. W. Stroke, and L. Rosen, *Proc. IEEE* **55**, 80 (1967).

9. W. E. Kock, *Laser Focus* **Feb.** (1969).

10. *Ultrasonic imaging and holography*, in: *Medical, Sonar and Optical Applications*, Plenum Press, New York, 1974.

11. ERC research in holography, *Electr. News*, **Oct. 10**, 32 (1966).

12. W. E. Kock, *Internal NASA-ERC Rev.*

13. F. B. Rotz and A. A. Friesem, Holograms with non-pseudoscopic real images, *Appl. Phys. Lett.* **8**, 146 (1966).

14. W. E. Kock and J. Rendiero, Some curious properties of holograms, *Proc. IEEE* **53** 1787 (1965).

15. T. S. Huang, R. A. Becker, W. E. Kock, and J. Rendiero, Comments on "some curious properties of holograms," *Proc. IEEE* **54**, 716 (1966).

16. G. Lippmann, *J. Phys. Paris* **3**, 97 (1894).

17. G. Lippmann, *C. R. Acad. Sci.* **146**, 58 (1908).

18. Y. N. Denisyuk, *Dokl. Akad. Nauk SSSR* **144**, 1275 (1962); *Sov. Phys.—Dokl.* **7**, 543 (1962).

19. G. W. Stroke and A. Labeyrie, White light reconstruction of holographic images using the Lippmann–Bragg diffraction effect, *Phys. Lett.* **20**, 368–370 (1966).

20. W. E. Kock, *Proc. IEEE* **54**, 331 (1966).

21. W. E. Kock, *Proc. IEEE* **55**, 1107 (1967).

22. W. E. Kock, Hologram television system and method, U.S. patent 3,548,093, Dec. 15, 1970.

23. C. W. Sherwin, J. P. Ruina, and R. D. Rawcliffe, Some early developments in synthetic aperture radar systems, *IRE Trans. Mil. Electron.* **Apr.**, 111–115 (1962).

24. L. J. Cutrona, E. N. Leith, L. J. Porcello, and W. E. Vivian, "On the application of coherent optical processing techniques in synthetic aperture radar," *Proc. IEEE* **54**, 1026–1032 (1966).

6

LASER FUNDAMENTALS

The earliest stimulated-emission device to be developed was the microwave maser; it was demonstrated in 1954.[1] For its development, the Nobel Prize in Physics was awarded jointly in 1964 to the U.S. scientist, Charles H. Townes, then at the Massachusetts Institute of Technology in Cambridge, Massachusetts, and to the Soviet scientists A. M. Prokhorov and N. Basov, both at the Lebedev Institute in Moscow. Figure 6.1 is a photograph of Professor Prokhorov and the author and was taken in 1959 in Prokhorov's laboratory in the Lebedev Institute. Following the success of the maser, many workers endeavored to extend its use from microwaves to light wavelengths. In 1960, the U.S. scientist T. H. Maiman, then at the Research Laboratories of the Hughes Aircraft Company in California, demonstrated the first laser, using a ruby rod as the active element.[2] His original laser is shown in Figure 6.2. Since the basic process for generating light is common to all three types of lasers, let us examine the workings of Maiman's ruby laser in some detail.

The active material, which in this form of laser can be ruby or various kinds of especially doped glass, is shaped into a cylindrical rod, as shown in Figure 6.3. Around this rod is wrapped a helical flash tube which, when connected to a powerful source of stored electrical energy, emits a very short and very intense burst of broad-band (incoherent) light. Some of this light energy is absorbed by the atoms of the rod, and in this process the atoms are excited, i.e., they are placed in an *energy state* which is at a higher *energy level* than the state in which most of them reside. Energy is thereby stored in these atoms, and when they return to their normal *un*excited state (we shall call this the *ground state*), they release their stored energy in the form of light waves.

Up to this point there is nothing special about the light-emitting process

FIGURE 6.1. A. M. Prokhorov, one of the three joint recipients of the Nobel Prize in Physics in 1964 for contributions to the field of masers, entertains the author (left) in his laboratory of the Lebedev Institute in Moscow.

just described; practically all light sources generate their light in this way. Thus, the atoms in an incandescent lamp filament are excited to higher energy states by the heat energy provided by the electric current passing through the filament, and as these atoms return to lower energy states (only to be again excited), they release this energy difference in the form of light.

The Danish scientist Niels Bohr suggested many years ago that the radiation of spectral lines by atoms could be explained by assuming that electrons revolve about the atom's nucleus in certain fixed orbits (like the planets circle the sun) and that each of these orbits represents a definite energy level. When an electron is in an outer orbit the atom is in a state of higher energy, an excited state, and when the electron transits to an inner orbit, energy is radiated in the form of spectral lines that are characteristic of a particular atom. Bohr thereby obtained values for the spectral series of visible hydrogen lines (colors) with an accuracy which was quite astonishing.

The more sophisticated theories of today show that these "orbits" are

FIGURE 6.2. A photograph of the first laser. It was designed by T. H. Maiman at the Hughes Aircraft Company.

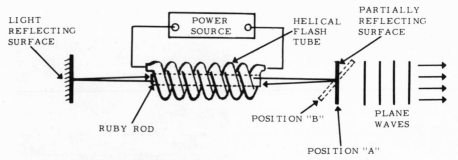

FIGURE 6.3. The essential parts of Maiman's ruby laser.

regions where the electrons are most likely to be. Albert Einstein showed that the emitted or absorbed light is exactly proportional to the change in energy and therefore to the change in "orbits."

6.1. The Metastable State

The significant difference between the laser and other light sources is that the laser light-source material provides a particular form of energy state in which the excited atoms can and do pause before returning to their ground state. They tend to remain in this state (called a *metastable* state) until stimulated into returning to the ground state. In this last step, they emit light having exactly the same wavelength as the light which triggered them into leaving that state. The atoms are thus *stimulated* into *emitting*; hence, the words "stimulated emission" in the laser acronym. Stated another way, energy is first stored in the atom and later released by it, when it transfers from the metastable state to the lower one, in the form of single-wavelength light energy.

6.2. The Two-Step Process

In the laser process, the wavelengths of the emitted light waves are exactly the same as the wavelengths of those which stimulated the atom to emit; these new waves are thus exactly suited to react with *other* metastable atoms and cause them to emit more of this same radiation. The usual laser process is not a one-step process, it is instead a two-step process. This is indicated in the energy diagram of Figure 6.4. Such energy diagrams, representing atomic processes, are plotted vertically, with the lower energy states placed low on the vertical scale, in parallel with the situation where objects at lower heights have lower (potential) energy. The lowest level, the ground state, represents that energy level to which atoms in the metastable state transfer. The excitation energy of the flash tube imparts to atoms residing in the lowest level sufficient energy to raise them to the energy state represented by the highest of the three levels shown. From this level they fall to the middle-level state, the metastable state. As we noted, the nature of this metastable level is such that the atoms tend to remain in it. When, however, the exactly correct wavelength light impinges on an atom in this state, it will depart from this state and fall to the lower energy state, emitting in the process a burst of energy in the form of light. The light burst which accompanies this return to the lower level can cause further emission of the same wavelength radiation from other atoms residing in the metastable state.

We see, therefore, that laser action depends upon the existence of this special state in the laser material. The process of placing, by means of the high-energy flash lamp, a large percentage of atoms in the laser material in the metastable state is referred to as a *population inversion*, because, originally most of the atoms reside in lower energy states. After the flash, an increased number of atoms reside in the metastable state.

Just how extraordinary can a metastable state be? As a general rule, the time that an atom spends in a normal excited state (this time is referred to as the *lifetime of the excited state*) is on the order of a hundred millionth part of a second (10^{-8} sec). In contrast, there is one metastable state which has an average lifetime of almost a full second (100 million times longer than that of the average excited state). This state is involved in the generation of the green line of atomic oxygen (at a wavelength of approximately 22 millionths of an

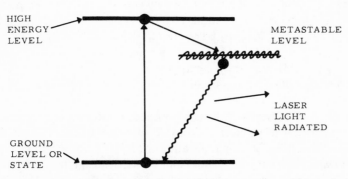

FIGURE 6.4. Three energy levels are available to atoms or ions in a laser material; transition from the middle, metastable, state to the ground state results in the emission of laser light.

inch), which is observed in the luminous night-sky phenomenon called the *aurora*. (The name *aurora borealis*, meaning "northern dawn," was given to this phenomenon in 1620 by the French philosopher Pierre Gassendi.)

6.3. Energy Conservation with Reflectors

The freely radiating light waves which issue from an atom which has been stimulated to emit radiation would normally travel outward in all directions. The use of *reflecting* surfaces in lasers helps to conserve the desired light-wave energy. Such surfaces cause light-wave energy to be reflected back toward the source, and this energy can then continue to stimulate more atoms into emitting more of the special radiation.

This use in a laser of reflecting surfaces to conserve the desired light energy is shown in Figure 6.3. Some of the light generated by metastable atoms returning to their lower energy state will travel down the ruby rod and be reflected by the reflecting surface, shown in Figure 6.3 as separated from the rod itself. (In some early ruby lasers, the ends of the rod were coated, and these ends acted as the reflecting surfaces.) Much of this light will be reflected back into the rod, and it can thus stimulate other atoms into emitting more of the desired light energy. The stimulated-emission process thereby builds up, and the result is that huge numbers of atoms soon radiate one powerful pulse of laser light.

Unless the conservation process of reflection is provided, lasers often are unable to oscillate, i.e., to generate light. For many of the most important applications of the microwave maser, on the other hand, the oscillating or wave-radiating condition is not desired. Masers are used primarily for amplifying weak radio signals, and therefore only enough stimulated emissions are permitted to provide amplification, and not enough to cause uncontrolled oscillation. For lasers, the greatest interest at the present time is in their light-wave-generation capabilities. However, in some applications, as we shall see, the equally possible *amplifying* capability of the laser is important. A laser pulse can be amplified by a laser amplifier to give it more energy, and many laboratories are exploring the possibilities of light waves propagating in tubes or glass fibers as a means of transmitting signal information (just as radio waves propagating in cylindrical, coaxial, cables now carry telephone and television signals). In such an application the light beams would be varied in intensity (*modulated*) at an extremely high rate, with the information being *coded* (as a telegraph signal is now coded, but at a very low rate). This modulation procedure could permit the transmission, via light beam, of millions of simultaneous telephone conversations and many thousands of simultaneous television signals. Because the energy of such a modulated light beam would decay in intensity with distance, the amplification capability of the laser could permit the signal carrying light beam to be amplified back up to its original strength.

6.4. Reflectors and Spatial Coherence

In addition to conserving energy, the reflectors also enhance the spatial coherence of the laser light since good wave planarity is equivalent to good spatial coherence. In the case of the sound waves in Figure 6.5, this planarity is achieved easily with a lens because the wave *source* is very small, the small horn having an aperture of only a wavelength or so in size. Because such small sources are not possible at light wavelengths, good planarity for light waves must be achieved in some other way. In the laser, the reflectors provide the means. When the single-frequency waves generated by the laser are made to bounce back and forth between two accurately made reflecting surfaces, they acquire, even after a few traverses, a very high degree of planarity. In Figure 6.3, one of the reflectors (the one on the right) is made slightly transparent so that not all of the energy striking it is reflected. Thus, that

FIGURE 6.5. An acoustic lens converts into plane waves the circular wave fronts of sound waves issuing from a horn placed at the focal point of the lens.

light which does not pass through it is the radiated laser light, available for use.

6.5. Energy Concentration

In Figure 6.3 the light waves emitted at the right from the laser are shown as a progression of plane waves having high *spatial* coherence; in addition, the nature of the light generation ensures that these waves have high *frequency* coherence. Having this full coherence, these waves can be manipulated in exactly the same way that coherent sound waves or coherent microwaves can be manipulated. For example, if a lens is placed immediately to the right of the plane-wave area at the right of Figure 6.3, it will focus the highly coherent, plane laser waves into an extremely small volume. This focusing of the laser light by a lens is equivalent to the lens action of Figure 6.5 reversed; in that figure, the waves *originate* at the small horn aperture at the left. However, one could assume that the plane sound waves at the right are arriving *from* the right. In progressing leftward through the lens, these plane waves would be converted into spherical waves and concentrated into a small volume comparable to the dimensions of the horn aperture (only a wavelength or two across).

With an optical lens placed at the right of Figure 6.3 the laser light is similarly concentrated into a volume perhaps only a wavelength or two across, and because light waves are only a few millionths of an inch in length, the power flow per unit area in this focal region can become fantastically high. An idea of the enormity of this energy concentration can be gotten by calculating what the power flow would be if all of the energy radiated by a 75-W light bulb were caused to pass through a square area one wavelength of violet light (16 millionths of an inch) on a side. This would be equivalent to a power flow per square *inch*, of approximately 300,000 million watts, i.e., equivalent to a funneling of this enormous amount of power through a 1-square-inch area. This amount of *electric* power is larger than that which could, in 1967, be generated by *all* the electric power stations in the entire United States operating simultaneously (this generating ability was 260,000 million watts). Even the very first lasers could make manifest their tremendous energy concentration capability in numerous spectacular ways. For example, laser light, when sharply concentrated, can punch holes in steel razor blades, as is shown in Figure 6.6; the flying sparks indicate that a hole is being made in a razor blade. Even without focusing, the beams from present-day truly powerful lasers can create rather spectacular effects just in the air through which the beam propagates, as shown in Figure 6.7. Figure 6.8 shows another example of the ability of the beam of a laser to travel great distances and still display its brightness.

FIGURE 6.6. A steel razor blade disintegrates when the coherent light waves from a ruby laser are focused sharply on it (courtesy of Bendix Aerospace Systems Division).

FIGURE 6.7. Even without focusing, the beams from present-day truly powerful lasers can create rather spectacular effects.(Courtesy of Hadron, Inc.)

6.6. *Q*-Spoiling

We mentioned earlier that lasers generally do not oscillate (emit light) when the reflectors are absent. In addition, a laser will usually not oscillate unless the two reflectors are positioned properly. This situation had made possible the obtaining of extremely powerful laser pulses. If, in Figure 6.3, the right-hand reflector is tilted (as shown by the dotted lines) so that it no longer reflects energy back into the laser rod, the normally desired energy conservation process is negated. However, the flash-lamp pulse, during its very brief lifetime, continues to place more and more atoms into their metastable state. Because very few are stimulated into falling back to the ground state, a much larger population inversion is achieved; many atoms are elevated to their metastable state, and few can leave it. If, now, at the proper instant, the reflector is rotated into its *proper* energy conserving condition, a very sudden, almost avalanchelike, transition of atoms takes place from the metastable state to the ground state, and a very short and extremely powerful pulse of single-frequency light is generated.

This process of negating the energy reflection or conservation procedure in lasers is called "*Q*-spoiling." The term arose from the expression used in radio engineering. In some radio sets, the tuning circuit comprises an inductance and a condenser, and the quality, or sharpness of tuning, of this circuit is specified by the letter *Q*, the first letter of the word quality. A sharply tuned (high-*Q*) circuit passes a very narrow band of frequency (it could, for example, in Figure 2.4, cause the 1% band shown to become appreciably narrower).

This higher quality is usually achieved by reducing circuit losses. If the losses are too high, that is, if the Q of the circuit is too low, a radio *oscillator* employing such a tuned circuit cannot oscillate. In direct analogy with this inability of a radio oscillator to oscillate because of the poor Q in its tuning circuit, the tilting of a laser mirror so as to inhibit the laser from oscillating is now referred to as *Q-spoiling*. When the mirror becomes properly positioned (so that the Q of the laser circuit is no longer "spoiled") oscillation occurs.

FIGURE 6.8. Partially illuminated by the sun off to the right, the earth appears as a crescent to a U.S. moon probe, NASA's Surveyor VII. Directly above the arrow are two tiny spots of light formed by laser beams aimed toward the moon and originating at Kitt Peak, Arizona and Table Mountain, California (courtesy of NASA).

This effect will become clearer when the analogy of the laser reflectors to acoustic or microwave cavity resonators is discussed.

6.7. Gas Lasers

The second form of laser to be developed was the gas laser.[3] It consists most simply of a glass tube filled with a special gas mixture. High voltage is applied across two electrodes near the ends of the tube, as shown in Figure 6.9, causing an electrical discharge to take place. The gas glows and the tube looks much like the glass tube of an ordinary neon advertising sign.

The gas laser differs from a neon sign in that its gas mixture provides the necessary metastable state in which the excited atoms can temporarily reside. As in the ruby laser, the energy difference between the metastable state and the laser state to which they fall corresponds to the energy of the single-color light which is radiated. The first gas laser used a mixture of helium and neon, and, like the first ruby laser, generated red light; its light had, however, an orange-red color, rather than the crimson red of the ruby. Its wavelength was 63.3 millionths of a centimeter (6328 Å), whereas the light from the ruby laser has a wavelength of 69.3 millionths of a centimeter (6930 Å). Other gases are now used: argon providing blue-green (4880 Å) and green (5145 Å) laser light, nitrogen providing ultraviolet (3371 Å) laser light, and carbon dioxide providing infrared laser light (106,000 Å).

One difference between gas lasers and ruby lasers is that most gas lasers operate continuously. The glow discharge caused by the applied voltage continually places a large population of atoms in a metastable state, and although many atoms are constantly falling back to a lower level, many are continually elevated again by the glow-discharge phenomenon.

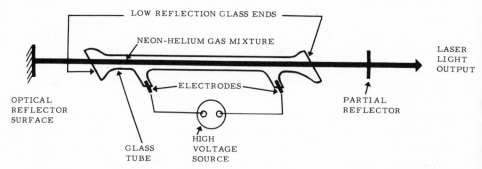

FIGURE 6.9. Gas lasers often use a direct-current glow discharge for exciting the atoms or ions to the required higher energy levels.

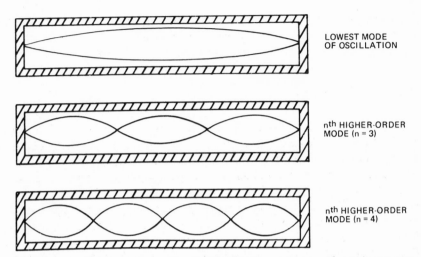

FIGURE 6.10. In a long box with closed reflecting ends, sound waves can resonate at various frequencies, as in an organ pipe.

6.8. Reflectors and Resonators

The ends of the glass tube of a gas laser are usually fitted with special low-reflecting glass surfaces as shown in Figure 6.9. The tube is terminated at each end with optically flat glass plates oriented at Brewster's angle* to eliminate reflections (for one polarization). The needed reflecting surfaces are placed externally to these, as shown. For most gas lasers the active portion (the gas tube) is much longer than the ruby rod of a pulsed laser. This feature, along with the feature of continuous operation, significantly improves the coherence of their light. The two reflectors in a laser act like the end walls of a resonator, and it is this greater distance between reflectors which enhances their frequency coherence. To clarify this point, let us review certain properties of resonators.

A piano string is a resonant device. It can vibrate either at its fundamental frequency or at any multiple of that fundamental. In like manner, a closed box (or resonator) can vibrate or resonate at its lowest frequency, or at any one of its higher-frequency or higher-order *modes*. This is illustrated in Figure 6.10. Here the box is small in cross section and resembles the resonator of a wooden organ pipe. At the top of the figure is sketched the wave pattern of the lowest or fundamental mode of oscillation. Reflection occurs at each end and maximum air motion exists at the center. In the center is sketched the third-order mode; this oscillation has a frequency three times that of the

---

* Brewster's angle is that angle of incidence for which the sum of the angles of incidence and refraction equal 90°.

oscillation at the top; it is said to be its third harmonic. At the bottom is sketched the fourth-order ($n = 4$) mode; it has one-fourth the wavelength and four times the frequency of the fundamental. For this wave, the length of the resonator is seen to be four half wavelengths (or two full wavelengths).

Let us imagine the four side walls of the wooden resonator to be removed and its two reflecting ends replaced with accurately made mirrors placed 8 inches apart. The separation between them would then be, for violet light (whose wavelength is 16 millionths of an inch), one million half wavelengths. Accordingly, if violet light is resonating between these two mirrors, a change to the next resonant mode, up or down, would result not in the 3 to 4 change of Figure 6.10, but rather in a far smaller percentage change, a change of only one part in a million.

In Figure 6.11, a frequency plot is drawn for the three cases of Figure 6.10. The two center drawings represent the higher modes corresponding to $n = 3$ and $n = 4$. In the bottom sketch an extremely large increase in frequency is indicated by the break shown at the left in the horizontal frequency scale. That sketch is intended to represent a condition similar to that of the two-mirror violet-light resonator just discussed; let us assume that each frequency indicated differs from its adjacent one by one part in a million. If the violet light reflecting back and forth between these two reflectors could be forced to resonate in only one mode, its single-frequency qualities (its fre-

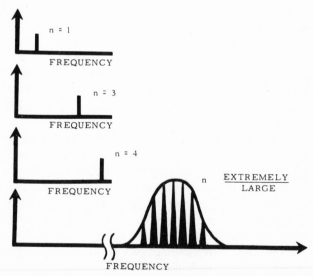

FIGURE 6.11. The top three sketches are frequency analyses of the three wave resonances of Figure 6.10. The fourth portrays the extremely high-frequency light-wave resonances which are possible with widely spaced laser reflectors.

quency coherence) could be described as having a constancy or stability significantly larger than one part in a million.

6.9. The Natural Line Width

A smooth curve is also indicated in Figure 6.11. We stated earlier that the natural atomic process by which laser light is generated is an important factor in providing the single-frequency characteristic. However, even this constant-energy process does not produce exactly single-frequency light. The observed line width is *broadened* by several effects, one is Doppler broadening and another is relaxation broadening; both are traceable to the random motion of the emitting atoms. Thus, although the energy process itself corresponds to an extremely narrow spectral line, the radiated line we observe has a significant frequency spread; this width is usually referred to as the *natural line width*. The smooth curve or envelope of Figure 6.11 is intended to indicate this natural line width.

We recall that the Doppler effect is noticeable as a drop in frequency or pitch of the whistle of a fast locomotive as it passes an observer. As the source approaches the observer, the sound possesses an "up-Doppler." To the locomotive *engineer*, however, no change is observed. For a moving atom, the frequency of its radiated light may similarly be quite constant and precise, *relative to the atom*, but to an observer located along the line of its motion, the light frequency would be Doppler-shifted.

An interesting use of artificially imparted Doppler shift to laser light was suggested by the U.S. scientist, Richard Milburn at Tufts University.[4] In his procedure, laser light is reflected from a beam of rapidly moving (high-energy) electrons issuing from an electron accelerator. Just as a slowly moving billiard ball has its direction reversed and its velocity (energy) greatly increased when struck by a second, oncoming, rapidly moving, billiard ball, so too do the reflected laser light waves acquire a very large Doppler increase in their frequency. Higher-*frequency* light means higher-*energy* light, and in this process the light is shifted so drastically in frequency that the laser beam is transformed into a beam of high-energy x rays (actually gamma rays). Furthermore, these gamma rays have properties virtually impossible to obtain with ordinary gamma-ray sources. First, the laser light is, or can be, *polarized* (as by passing it through the glass from a pair of Polaroid sun glasses), which causes the resulting *gamma* rays *also* to be polarized (a very unusual situation). Second, the monochromaticity of the laser light (high frequency coherence) results in the gamma radiation similarly being quite monochromatic (this feature is also quite unusual).

Recently some atomic transitions have been observed for which the natural line width is quite narrow. One procedure for achieving this was discovered and utilized by the German scientist, Rudolf Mössbauer [5,6] in 1957 (he received the Nobel prize for his work in 1961). He employed a radiating substance in which the atoms were so firmly bound in a crystal lattice that they could not move, even when they radiated. The elimination of motion in these atoms caused their radiation to be extremely frequency coherent. Unfortunately, this procedure cannot be used in lasers; for them, a sizeable natural line width is always observed.

Fortunately, however, the individual resonator modes of a laser can be appreciably narrower than the natural line width of the laser, and hence this narrower width provides a significant improvement in frequency coherence. There often is a problem in keeping the oscillation in only *one* of the numerous modes shown in the lower part of Figure 6.11. Many lasers oscillate in several modes simultaneously, and for others, mode "jumping" is constantly occurring. These phenomena, obviously, impair the coherence of the laser light. To minimize the mode-jumping effects, extremely high mechanical stability is usually employed. Thermal contraction of the reflector supports can result in a shortening of the distance between reflectors, thereby causing the oscillator frequency to change from one resonator mode to another. When, however, extremely good mechanical and thermal stability is maintained, it is possible for the laser to operate in one single mode of oscillation and thereby possess an extreme degree of frequency coherence.

6.10. Coherence Length

Because gas lasers provide excellent coherence, they are the lasers most commonly used in making holograms. But even here some gas lasers are better than others. For holography the most pertinent way of defining the extent of a laser's frequency coherence is by specifying its *coherence length*. Let us examine what this term implies.

We noted earlier that interference effects still occur for single-wavelength waves, even when one of the wave sets is slid an integral number of wavelengths ahead or behind the other set. This, of course, assumes that the separation between wave crests remains identical and that each wave crest and trough is exactly like all succeeding crests and troughs. Actually, we find that after one set has been shifted relative to the other by many, many wavelengths, a point is eventually reached where the addition and subtraction effect is no longer perfect. It is easy to see that this is a real problem at light wavelengths. When two wave sets of plane, coherent, violet light waves are shifted only 16 inches, this shift is one encompassing one million wavelengths. A wavering in frequency of only one part in a million would cause these two crests to be out of step.

The distance over which laser radiation *does* remain in accurate step is called the coherence length; some gas lasers exhibit coherence lengths as large as several meters. For use in holography, lasers having such long coherence lengths permit the recording of scenes having much greater depths than could be recorded with shorter coherence-length lasers.

6.11. Semiconductor Lasers

Another form of laser to be developed fairly early was the semiconductor, or injection, laser. It is similar to the ruby and gas laser except that it employs as its active substance a tiny piece of semiconductor material. A semiconductor is a material which is neither a good conductor of electricity nor a perfect insulator. When used as a laser, direct current is made to flow through the material by connecting a power source to two electrodes affixed to opposite surfaces. Atoms within the material are thereby excited to higher energy states, and in falling back to a lower state, give off light or other forms of electromagnetic radiation. Solid-state lasers are particularly useful for generating radiation in the infrared region and in the millimeter radio-wave region. They have not been used appreciably up to now in holography.

Laser action has been produced in certain substances (including liquids) by utilizing a light-scattering process called, after its discoverer, the Indian scientist Sir C. V. Raman, the Raman effect. This effect is observed when light waves falling on a material interact with the internal vibrations characteristic of the material, thereby causing sum-and-difference-frequency light waves to be generated. Raman received the Nobel Prize in Physics for this effect in 1930. Figure 6.12, taken at the author's home, includes Sir C. V. Raman and two other Nobel Laureates.

In some arrangements, a resonator is made of the Raman material (such as, for example, a block of quartz having plane, parallel end faces[7]) and a powerful ruby laser beam serves as the means for exciting the *lattice* vibrations in the material. Thus, just as the flash tube is the "pump," or energy source, for the ruby laser, a laser beam here acts as the "pump." The intense lattice vibrations generated in the quartz *modulate* the ruby light beam, thereby generating the upper and lower *sidebands* (sum and difference frequencies). In parallel with the term stimulated emission, this process[8] is referred to as "stimulated Raman scattering" (SRS).

FIGURE 6.12. The discoverer of the effect utilized in some lasers, Nobel Laureate Sir C. V. Raman (standing center) poses at the author's home with a group of Bell Laboratories scientists. Among them are Nobel Laureates Walter Brattain (seated beneath Raman) and John Bardeen (second from left, standing).

6.12. Other-Level Lasers

We have limited our discussion to the 3-level case of Figure 6.4. There are many materials in which more than three energy levels are involved when operating in a laser. As the author has noted,[9] for laser specialists thinking of multilevel transition models, it is refreshing to recall the fact that C. H. Townes' original maser operated successfully because he was able to separate *two* molecular levels of ammonia molecules by an inhomogeneous electrical field, and to lead the higher-energy molecules into a resonant microwave cavity. The frequency of resonance, ν, of the cavity was related to an energy $h\nu$ (the energy difference between the two levels). The higher-energy-level set of molecules could thus be stimulated (by a weak, incoming, received radio signal) into emitting radiation of a frequency ν (23,870 MHz), thereby causing the incoming radio signal to be amplified. It is to be noted that this scheme was conceived, and the first maser built, by Townes and his students at Columbia University in 1954.[1]

6.13. References

1. J. P. Gordon, H. J. Zeiger, and C. H. Townes, The maser—new type of amplifier, frequency standard, and spectrometer, *Phys. Rev.* **99**, 1264–1274 (1955).
2. T. H. Maiman, Stimulated optical radiation in ruby, *Nature* (London) **187**, 493–494 (1960).
3. A. Javan, W. R. Bennett, Jr., and D. R. Herriot, Population inversion and continuous optical maser oscillation in a gas discharge containing a He–Ne mixture, *Phys. Rev. Lett.* **6**, 106–110 (1961).
4. R. H. Milburn, Polarized gamma rays from back-scattered laser light, paper presented at the February 1968 Meeting of the American Physical Society, session G. B.
5. R. L. Mössbauer, *Z. Phys.* **151**, 124 (1958).
6. W. E. Kock, The Mössbauer radiation, *Science* **131**, 1588–1590 (1960).
7. J. P. Scott and S. P. S. Porto, Longitudinal and transverse optical lattice vibrations in quartz, *Phys. Rev.* **161**, 903 (1967).
8. D. vonder Linde, M. Maier, and W. Kaiser, Quantitative investigations of the stimulated Raman effect using subnanosecond light pulses, *Phys. Rev.* **178**, 11 (1969).
9. W. E. Kock, Review of "Lasers," by B. A. Lengyel, 2nd ed., Wiley, 1971, *Appl. Opt.* **11**, 218–219 (1972).

7

RECENT DEVELOPMENTS IN LASERS

Since the early work of Maiman, a long list of new types of lasers has appeared. Here we will review those developments which have had particular impact upon the engineering applications possibilities of lasers and holography. Because we shall be discussing how "powerful" some of the recent lasers are, we first recall (1) the various units of power, energy, and dimension (in space and time), and (2) the ranges of frequency and wavelength usually associated with the ultraviolet, visible and infrared regions of the spectrum.

7.1. Laser Units

A watt (W) is the equivalent of 10^7 ergs/sec, and a joule (J) is equivalent to 10^7 ergs or 2.778×10^{-4} watt-hours (w-h). One watt is therefore equal to one joule per second. These units are important when discussing the maximum power which a laser is capable of generating, or the energy (in joules) associated with the pulse length of the laser. From the above we see that if the pulse has the rather long length of one second, an energy in the pulse of one joule would be equivalent to a maximum power of only 1 watt. If, however, the pulse has a length of 10^{-6} sec and has an energy of 1 J, the maximum power in the pulse is 10^6 W (1 MW), since the maximum power in watts times the pulse duration in seconds equals the energy per pulse in joules.

When the laser pulses are repetitive, one can speak of an average power output of a laser (in watts), this being the product of the energy per pulse (joules) and the pulse repetition rate (in Hz, i.e. pulses per second). Thus, pulses having 1 joule of energy, if repeated 10 times per second (10 Hz), have an average power of 10 W. The basic unit of time is of course the second, a millisecond (msec) being 10^{-3} sec, a microsecond (μsec) being 10^{-6} sec, a nanosecond (nsec) being 10^{-9} sec, and a picosecond (psec) being 10^{-12} sec. Going to positive exponents, for example, with Hertz (cycles per second), a kilohertz (kHz) is 10^3 Hz, a megahertz (MHz) is 10^6 Hz, a gigahertz (GHz) is 10^9 Hz, and a terahertz (THz) is 10^{12} Hz.

One of the common units of length (wavelength) used in optics is the Ångstrom (Å), which is equal to 10^{-10} m. However, the micrometer (μm) is also widely used; thus the wavelength of the helium–neon (He–Ne) laser radiation can be specified as being either 6328 Å or 0.6328 μm. In terms of micrometers, the ultraviolet region of the spectrum is often specified as the region 0.1–0.4 μm, the visible as 0.4–0.7 μm and the infrared as 0.7–100 μm.

7.2. Solid Lasers

One class of lasers earlier designated as solid-state is now usually referred to as solid. We shall employ the term solid because the expression solid state was originally applied in the semiconductor field following the extension of the earlier *surface* effects in World War II point-contact microwave diodes to *solid*-state junction devices (diodes and transistors). Many of the solid-state semi-conductor materials, which were originally investigated for diodes and transistor applications, have since been found to be effective as semiconductor *lasers*.

Maiman's laser with its rod of ruby, falls in the solid-laser category. The rod of the laser is fashioned from a grown ruby crystal and, as in other solid lasers, the bulk of the rod acts merely as the "host" material, with the "active" material being the chromium ions. The composition of ruby is represented as $Cr^{3+}:Al_2O_3$; when the material corundum (aluminum oxide, Al_2O_3) possesses no chromium, it is called sapphire. When the coiled flash lamp of Figures 6.2 or 6.3 is triggered, the chromium ions in the ruby absorb the green and blue light of the flash lamp by virtue of their broad absorption bands, thereby raising many ions from their ground state to the highest of the three levels shown in Figure 6.4. (Actually, this upper level consists of several fairly broad levels.) From this region, the ions spontaneously transit to the lower, very narrow, metastable level, which then can become more densely occupied than the ground level, i.e., a population inversion can exist. In this situation, the ruby can act as an "amplifier" for light waves having a wavelength of 6943 Å.

The ruby laser is classed as a *crystalline* solid laser because the host material is a grown crystal of ruby. Other host crystals which have been found to permit lasing when certain active ions are incorporated are: yttrium aluminum garnet (YAG, $Y_3Al_3O_{12}$), calcium tungstenate ($CaWO_4$, W being the symbol for tungsten, i.e., wolfram), also calcium fluoride, strontium fluoride, magnesium fluoride, and zinc fluoride. Active ions for these hosts include chromium, cobalt, nickel, neodymium, trivalent uranium, and other rare earth ions. In this connection it is interesting that of all the crystalline lasers, Maiman's ruby laser is the only one involving three energy levels; all of the rest involve four levels. In these, the added level is rather close to the ground state, so that it is the transition from the high metastable state to this near-ground-state level which provides the radiated laser light. The reporting of the first four-level laser occurred in 1960 shortly after Maiman's discovery.[1]

Probably the most important of the more recent lasers is one using neodymium ions as the active ions and noncrystalline (amorphous) glass as the host material. This use of glass bypasses the cumbersome crystal-growing process, permitting larger rods to be used, with a consequent higher peak power. Stimulated emission in neodymium-doped glass was first reported in 1961 by E. Snitzer,[2] and since then developments in neodymium–glass lasers have mushroomed, with some lasers having rods 2 m in length and 3.8 cm in diameter. The excitation is usually a xenon flash tube and the peak of the radiation is around 1.06 μm. The powerful laser action shown in Figure 6.7 was accomplished by means of a neodymium–glass laser.[3] In that same 1971 reference, Solon compared laser powers with the power capacity of the U.S. electrical utility industry in 1971, which we mentioned in the last chapter was in 1967 under 300,000 million watts, and which by 1971 had risen to 335,000 million watts. He observed that "lasers are available at several laboratories in our country that routinely generate pulses of coherent light of 100 joules in one nanosecond" (an instantaneous power of 100,000 million watts), and "at least one laboratory has reported production of laser pulses of 20 joules in 10 picoseconds" (equivalent to an instantaneous power of 2,000,000 million watts). He notes that it is not the *energies* which are impressive (an ordinary 100-W light bulb consumes 100 J of electrical energy every second), but rather the *control* that "permits manipulation of short time intervals to yield such prodigious powers." Thus, 10 psec is the time taken by light waves (velocity 186,000 miles/sec.) to travel 3 mm (about the thickness of a U.S. nickel coin).

The large maximum powers we have mentioned are generally the result of a laser amplifier chain *amplifying* the output of a large laser, and we shall discuss solid-laser amplifiers (including disc lasers) in Section 7.3. Before this however, we briefly note that lasers using neodymium (Nd) as the active element and yttrium aluminum garnet (YAG) as the host material have been quite successful operating either in the continuous-wave (CW) condition or in the Q-switched pulsed mode. Several sizes are available commercially for use in machining, scribing, and resistor trimming. In one experiment using such a Nd–YAG laser, a grown crystal of barium sodium niobate ($NaNb_5O_{15}$) was placed in the laser cavity. The crystal acted as a harmonic generator, causing coherent *green* light at 0.53 μm to be radiated.

7.3. Solid Laser Amplifiers

We mentioned in Chapter 6 that numerous laboratories are "exploring" the possibilities of laser amplifiers for communication systems employing light waves in contrast to present-day radio waves and microwaves. This activity followed the successful use of *masers* as amplifiers in many radio applications. At those wavelengths, the maser could be constructed as a single-mode device, and because of its high temporal coherence, it could act as an amplifier for very-narrow-frequency bandwidth signals. It also added a very low noise component and hence achieved a high signal-to-noise ratio.

In radio circuits, the disturbing noise is referred to as "first-circuit" noise, as it is generated in the *first* amplifying circuit of the system. For a given bandwidth requirement, B, and a temperature, t, at which the first amplifier operates, the intrinsic noise (noise which cannot be reduced) is proportional to KtB, where K is Boltzmann's constant. However, the first circuit, whether it be a simple detector or an amplifier, invariably adds a certain amount of "first-circuit noise," the extent of this added noise being specified by the "noise figure" of the circuit. Single-mode masers were found to have an exceptionally *low* noise figure and hence were useful in numerous applications.

A different situation exists for the laser. First, they are almost invariably multimode devices and, second, the number of spontaneous transitions (spontaneous emissions) compared to the desired stimulated emissions is always high compared to the lower-frequency maser. (This ratio increases as the cube of the frequency of the stimulated emission device.) A low-level signal, in need of amplification, tends to be drowned out by the noise generated by these spontaneous emissions.

Fortunately, the output of a pulsed solid laser is so high that it rides above the spontaneous-emission noise, and can therefore become amplified by a (solid) laser amplifier. The reason for using an amplifier in this case is simply to increase the maximum power in the pulse. Accordingly, use is often made of a series of amplifiers, each causing the pulse to be more powerful when it leaves than when it entered the amplifier (Figure 7.1).

The amplifier is similar in some respects to the pulse-initiating laser. It employs a rod of the same material (e.g., for a neodymium–glass laser, the amplifier rod would also be made of neodymium-doped glass) and it employs a similar flash lamp (e.g., a xenon lamp). However, when a number of amplifiers are employed, the successive stages usually employ rods of increasing diameter to keep energy densities below the damage threshold of the glass.[4] One series of rod diameters proceeds from 16 to 23 to 32 to 45 to 64 mm. Also, the ends of the amplifier rods are usually cut on the bias, i.e., at Brewster's angle, to reduce reflections at the rod ends so that the amplifier will not oscillate. Of course the laser "oscillator" is equipped with a Q-spoiling mechanism so that the original pulse is as large as possible to begin with. Isolators (see Chapter 1) are often placed between successive amplifier stages. Also, a device often employed is a pulse-sharpener. Such a unit, when placed in the path of a Q-switched laser whose pulse has a duration of, say, 20–100 nsec, can shut off the beam during the rise of the laser pulse, open the beam path near the peak of the laser pulse, and shut the beam off again at

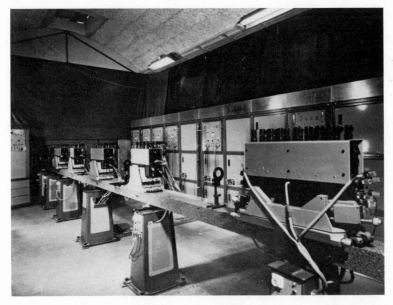

FIGURE 7.1. A chain of high-power laser amplifiers (courtesy of Hadron, Inc.)

an adjustable, preset, instant. The emerging light pulse is thus given a fast rise time and a controlled duration.

Because rod-type amplifiers, if overloaded, often experience such severe temperature gradients that fracture occurs because of the thermally induced stress, a device referred to as the disk laser (or amplifier) has received increasing attention in recent years. Figure 7.2 shows a 1969 tilted-disk laser amplifier constructed at the Lawrence Livermore Laboratory of the U.S. Atomic Energy Commission.[5] The 15 neodymium-doped glass disks are tilted at Brewster's angle to minimize reflections, and the open spaces between the disks prevent the large temperature buildup which occurs in a solid rod. More recently, a Livermore laser module design was reported[6] having four disk amplifiers, with disk diameters of 5, 9, 18, and 36 cm, and which is expected to generate 1000 J. The advantages of the disk design for moderate power lasers has been discussed,[7] particularly in regard to its ability to withstand high pulse-repetition rates. Another form which evolved from the disk laser is the so-called zigzag laser (Figure 7.3), in which each successive laser slab of neodymium–glass is tilted at the opposite Brewster's angle. Superior cooling properties are claimed for this arrangement.[8]

7.4. Gas Lasers

Following shortly on the heels of Maiman's ruby laser was the helium–neon gas laser, reported by Javan *et al.* at the Bell Telephone Laboratories in

FIGURE 7.2. A tilted-disk laser amplifier (Lawrence Livermore Laboratory).

FIGURE 7.3. The zigzag tilted-disk laser (courtesy of General Electric Company).

1961 (see Chapter 3, Reference 6). This class has truly mushroomed since that time, with a 1972 estimate indicating more than 2000 types,[9] covering the spectrum from ultraviolet (0.15 μm) to infrared (699.5 μm). Gas lasers are usually divided into three subgroups, based on their gain mechanism: atomic (e.g., helium–neon), molecular (e.g., carbon dioxide) and ionic (e.g., argon). The recent metal-vapor types are assigned to the ionic category.

Among those depending upon atomic transitions, the He–Ne was the first, and it is still the most widely used. As we indicated in Chapter 6 (Figure 6.9), a direct-current (dc) or radio-frequency (rf) glow discharge is usually employed for exciting the neon atoms. Although the helium (90%) becomes involved in some of the reactions, it is usually referred to as the host material, and neon the active material (10%). The He–Ne lasers (and other gas lasers) can be made very frequency stable, and because of this, they are the most nearly monochromatic sources (highest temporal coherence).

In contrast to the atomic gas lasers, which depend upon energy exchanges in atomic electron orbits, molecular gas lasers utilize energies associated with vibrations and rotations within a molecule. Because of this the energy-level structures in such lasers are much more complex than those in atomic lasers.

One of the most important of these is called the carbon dioxide laser, and although oscillations can be obtained in pure CO_2, far more powerful laser action is obtained when other gases such as nitrogen are added.[10] The early work on this laser was done by C. K. N. Patel at the Bell Telephone Laboratories (Figure 7.4), using a continuously flowing gas arrangement in the optical cavity. Nitrogen, carrying vibrational and rotational energy, mixes with the active gas (CO_2) and energy is transferred through collision. In these same experiments Patel demonstrated laser action in carbon monoxide (CO), nitrous oxide (N_2O), and carbon disulphide (CS_2). He found that the CO_2 laser was much more efficient than earlier gas lasers (up to 15% as compared to 0.1%) and that CO_2 is sufficiently stable chemically that the laser would function even if the discharge took place in the CO_2, thereby permitting the "flowing-gas" procedure to be bypassed.[11] Today there are many versions of sealed-off CO_2 lasers commercially available. Patel also found that the addition of helium increased the efficiency of the CO_2 laser. The waves generated by CO_2 lasers generally have a 10.6 μm wavelength (infrared). The early versions were CW.

An interesting variant of the original CO_2 laser is the TEA (Transversely Excited Atmospheric pressure) laser. The first experiments with transverse excitation occurred in 1965, when D. A. Leonard reported a transversely excited nitrogen laser.[12] In 1968, A. E. Hill reported the generation of multijoule *pulses* from a CO_2 laser,[13] and in March 1971, the U.S. Army Missile Command reported obtaining 286-J pulses from a CO_2 laser 9 m long and 15 cm in diameter.[14] In early 1970, J. A. Beaulieu and his associates at the Canadian Defense Research Establishment at Valcartier reported the

FIGURE 7.4. An early carbon dioxide laser (Bell Telephone Laboratories)

FIGURE 7.5. A recent, extremely large CO_2 laser (Wright Patterson Air Force Base).

development of the TEA laser;[15] it utilized many separate transverse discharges between separate pin cathodes and a common bar anode. The TEA lasers are characterized by their low cost and high efficiency, and have been found useful in industry for drilling, welding, cutting, and machining. More recently, scientists at the National Research Council of Canada developed a TEA laser capable of generating gigawatt powers in pulses 50 nsec in length. Smaller commercial models (for semiconductor scribing) have been available since 1971 (e.g., TRW, Inc.).

Other developments in CO_2 lasers include (1) a high-mass-flow technique in which the gaseous constituents (N_2, He, CO_2) are premixed and continuously recirculated axially, permitting 5–7 kW of power in a CW beam,[14] and (2) a very high-speed *transverse* flow of the CO_2 gas, leading to 10 kW of CW power in the beam.[16] The physical size of CO_2 lasers has grown fantastically over the years (Figure 7.5), and a new laser, the gas-dynamic (CO_2–N_2), bids fair to become larger still. These lasers utilize the thermal energy given off in a combustion process for pumping. Exhaust from the combustion involving air and a fuel containing carbon (the exhaust therefore comprising perhaps 10% CO_2 and 90% nitrogen) is allowed to expand supersonically through a nozzle. The metastable N_2 state does not cool as rapidly as the CO_2, and collision between the cooled CO_2 and the still-excited N_2 molecules provides the population inversion for the operation of the laser (see Reference 17; this reference includes a list of 128 references on CO_2 lasers). In a NASA sketch dated December 12, 1968, a gas-dynamic laser is indicated as putting out 1 MW of continuous laser power, with the "potential advantages" being shown as (1) power output significantly increased, (2) overall efficiency substantially increased, and (3) new realms of possible applications opened. Such lasers avoid the necessity of first converting thermal energy into electrical energy for operating the laser, and because of this similarity to power machinery (e.g., gasoline or diesel engines), the term "photon machines" has been applied to them.[17] There are reports that wooden planks have been ignited at a distance of 3 km with a CW gas-dynamic laser.[9]

Gas-dynamic lasers have also been operated in the pulsed mode,[18] and the gas-dynamic principle was applied to a mixture of carbon *mo*noxide (25%) and nitrogen (75%), with a CO laser as the result[19] having emissions in the 5-μm range (Figure 7.6). This development took place at NASA's Ames Research Center.

Several other recent gas lasers are of interest. A hydrogen fluoride (HF) laser, with its output at 2.6 μm (Figure 7.7) was reported in 1971.[20] In this

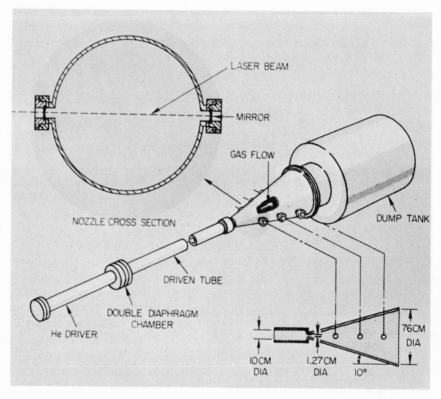

FIGURE 7.6. Arrangement for a CO laser (courtesy of NASA Ames Research Center).

FIGURE 7.7. A hydrogen fluoride laser having a wavelength of 2.6 μ.

version, the fluorine atoms were produced by an electrical discharge which decomposed sulfur hexafluoride. More recently, a *pulsed* HF laser was developed (Figure 7.8) in which the excitation source was the AEC's Sandia Laboratory's relativistic electron-beam accelerator (REBA). The accelerator pulse, a 55-kiloampere, 2-million-volt electron beam, produced a 228-J, 55-nsec pulse, equivalent to a maximum power of nearly 4 billion watts (Figure 7.9). The efficiency was reported to be about 8%.[21] In connection with the HF laser of Figure 7.7, it was noted that the effect could also have taken place by means of a chemical reaction instead of the electrical discharge. In a so-called *chemical* laser, the population inversion is produced as a direct product of a chemical reaction.

To date, one of the shortest-wavelength lasers is the molecular hydrogen

FIGURE 7.8. A *pulsed* HF laser using a relativistic electron-beam accelerator as an excitation source (courtesy of Sandia Laboratory).

one, first reported in 1972 by both IBM and the Naval Research Laboratory (NRL). The 0.116 μm vacuum ultraviolet radiation was achieved at IBM using a high-power electron beam as the pump source (Figure 7.10), and at NRL by means of a traveling-wave discharge.[22] Coherent ultraviolet radiation has also been generated using nonlinear materials [23] and a longer-wavelength powerful laser for generating second and fourth harmonics of the laser frequency (Figure 7.11). As for the long-wavelength region (the far infrared), scientists at the Aerospace Corporation (as reported in January 1974) utilized a *waveguide* cavity, modeled after a microwave cavity, and observed radiation at several wavelengths between 40 and 200 μm (using a CO_2 laser as a pump), with the strongest line being observed at 118μm (Figure 7.12).

Finally, in the category of gas lasers, the ion laser and the metal-vapor laser are recent additions. In an ion laser, stimulated emission occurs between two energy levels of an *ion*, rather than atom or molecule. The first materials to be used in such lasers were argon and krypton. The first argon ion lasers were set to oscillate at one frequency, 0.4880 μm. Today, in commercially available models, numerous radiation lines of the argon and krypton ion lasers can be chosen by adjusting a prism "wavelength selector." The argon

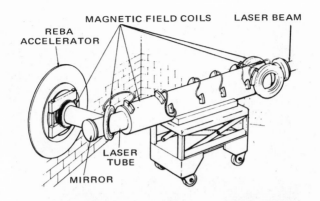

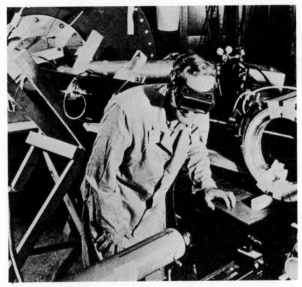

FIGURE 7.9. Details of the laser of Figure 7.8. This laser has a maximum power of almost 4×10^9 W.

laser lines range from 0.529 to 0.35 μm, with nine lines in between; the two which provide the largest power output have wavelengths of 0.488 and 0.515 μm. The krypton laser lines range from 0.75 μm to 0.476 μm, with the most powerful line being at 0.647 μm. Models are also available having a mixture of argon and krypton gas, permitting somewhat more power to be available in the areas where the two overlap. A more recent addition to the ion laser class is the helium–cadmium laser[24] (actually it is a "metal-vapor"

FIGURE 7.10. A molecular hydrogen laser, having a vacuum ultraviolet wavelength of 0.116 μ (courtesy of IBM).

laser), providing radiation at 0.44 μm (a "brilliant" blue) and at 0.325 μm (in the ultraviolet).

7.5. Semiconductor Lasers

In semiconductor injection lasers, electrical energy is converted directly into coherent light, with the conversion taking place by virtue of the work done by the impressed electric field on the charge carriers in the semiconductor material. To date, such a carrier-injection process is the most efficient procedure for the conversion of electrical energy into coherent radiation. Semiconductor lasers may also be excited optically, by electron bombardment, or by avalanche breakdown.

As in the case with other solid-state devices (Figure 7.13), semiconductor lasers are quite small compared to other varieties. The simplest variety resembles a solid state P–N junction diode, a device which was first developed around 1950 by William Shockley and his associates at the Bell Telephone Laboratories. In a *laser* diode, when electric current is passed through it, light is emitted at the junction area and it radiates out of the *side* of the

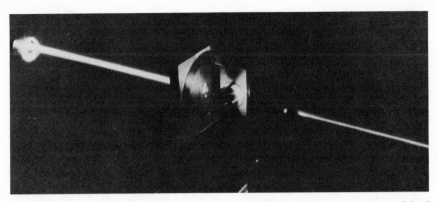

FIGURE 7.11. The production of coherent ultraviolet light can be accomplished by generating harmonics of a powerful lower-frequency laser.

junction. Even at low current levels, some light is emitted, but as the current is increased, a point is reached where the intensity increases rapidly, and the spectral width of the radiation decreases markedly, indicating the onset of stimulated emission.

Because of the heat produced by the large currents when lasing occurs,

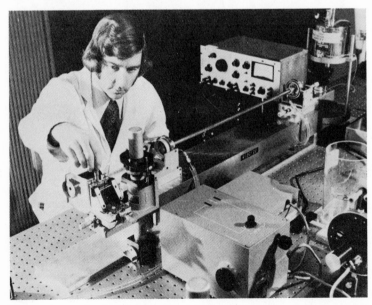

FIGURE 7.12. A far-infrared coherent source can be made using a waveguide cavity and a CO_2 laser as a pump (courtesy of Aerospace Corp.).

early devices had to be cooled to very low (liquid-nitrogen) temperatures. More recently, a technique has been developed for making the effective thickness of the diode junction much smaller (called a "heterojunction" laser) and thereby permitting the unit to be operated at room temperature. The semiconductor material most often used is gallium arsenide (GaAs); Figure 7.14 indicates how small such lasers can be.[25] A second variation in structure which permits larger power outputs uses a multiple-layer expitaxial process whereby 5 layers are formed, with the two which flank the central GaAs active region being aluminum gallium aluminum arsenide [$Al_xGa_{(1-x)}As$]. These are called "double-heterojunction" lasers. A variety called the "large optical cavity" laser employs a thin active (recombination) layer which is adjacent to a wider "cavity" or propagation layer, permitting larger power outputs.[26] A typical commercial model is shown in Figure 7.15.

Advantages of semiconductor lasers over other types include the ease of modulating the light output, including high-speed (pulse) modulation (typical

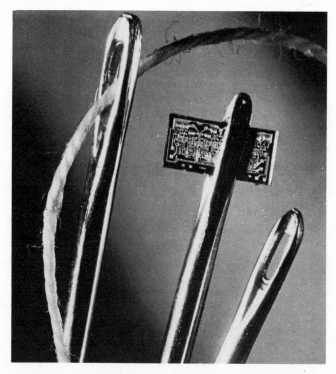

FIGURE 7.13. The extremely small components now possible in solid-state circuits (here shown in the eye of a needle) have led to very small semiconductor lasers.

FIGURE 7.14. A tiny gallium arsenide semiconductor laser placed on a U.S. coin (Bell Telephone Laboratories).

rise times are about 1 nsec), and, because there are very efficient solid-state detectors (e.g., silicon avalanche diodes) for the light wavelengths involved (0.9–0.66 μm), the use of these lasers in optical communication systems appears promising. Experiments at Bell Labs[27] utilized a bistable (Gunn-effect) switch for modulating the output of a double-heterojunction GaAs laser in pulse-code-modulation (PCM) experiments. Pulse rise times of 200 psec were observed, with 1.2 gigabit-per-second (Gbit/sec) operation stated as being possible.

7.6. Liquid Lasers

After early experimentation with various materials (e.g., rare earth chelates and the inorganic neodynium–selenium oxychloride), development in the liquid laser field has pretty well settled into the use of inorganic dyes of various types. One advantage of the liquid-dye lasers is their *tunability*,

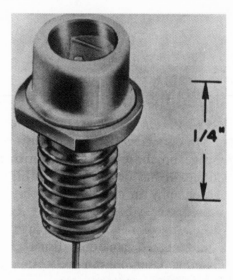

FIGURE 7.15. A "large optical cavity" semiconductor laser (courtesy of RCA).

permitting a wide range of colors of light to be generated. Thus, one of Bell Laboratories' dye lasers, called the exciplex (excited-state complex), can cover 1.760 μm, from the near ultraviolet to yellow region,[28] and the reaction occurs in organic dye materials (Figure 7.16). Tuning of these lasers is accomplished with a diffraction grating.

Recently the substance Rhodamine 6G, an organic dye that fluoresces in the red to near-infrared has been used both for single-wavelength radiation and for tuning over a (small) band (with a grating). Bandwidths of 1 Å can reportedly be obtained by this process.[29] Kodak, IBM, and Coherent Radiation are among the laboratories investigating this substance, with Coherent reporting an output of 9 W using an argon ion laser as a pump.[30] Kodak, in August, 1973, offered 10 different dyes for lasers, and Avco offered "desk-top" dye lasers, and "drop-in" dye "cassettes" for their (tunable) lasers.

7.7. Mode Locking

An effect achievable in lasers, and which created quite a stir when it was first discovered, is a process called mode locking. First demonstrated through the utilization of a high-frequency acoustic vibration of one reflector of a laser, the effect causes the (usually pulsed) laser to generate a series of extreme-ly sharp (short-duration) pulses, with an accordingly high maximum power generation in the pulses.

We saw earlier that numerous modes can exist within the natural line width of a stimulated emission reaction (Figure 6.11). For the huge number of wavelengths which exist in a standing wave existing between the two reflectors of a laser, as shown in Figure 6.9 (a distance of many inches with wavelengths of millionths of an inch), and because the frequencies of adjacent modes differ only by the slight amount equivalent to a one-wavelength change in the cavity length, the frequencies of all modes falling within the natural line width are *almost* identical. Accordingly, if a *frequency* modulator (FM) is imparted to the modes (by vibrating one of the reflectors of the optical cavity), and if this vibration frequency is approximately equal to the tiny frequency difference

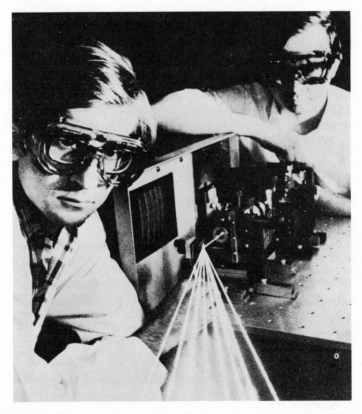

FIGURE 7.16. The many beams from this laser demonstrates the wide tunable range (courtesy of Bell Telephone Laboratories).

existing between two adjacent modes (actually among *all* adjacent modes within the line width), such a *small-index* FM yields only one pair of sidebands for each mode. Such side bands have opposite polarity (in FM diagrams the one to the left of the carrier is usually shown pointing downward and the one to the right upward), and, being spaced apart from the carrier by the modulation frequency (the mode separation), side bands and carrier frequencies of all modes fall on top of each other, with the resulting spectrum of the superposition of all (vibrating) modes (all frequencies) being a long series of single frequencies, all spaced apart at the frequency of the acoustic vibration. Such a multitude of equally spaced frequencies is exactly the spectrum of a series of repeated, extremely sharp, pulses of the "carrier" frequency ("carrier" is in quotes because the original mode frequencies are not *exactly* alike, but have become *locked* together in phase by the process). Such a series of short pulses has been suggested for use in several applications.

7.8. References

1. P. P. Sorokin and M. J. Stevenson, Stimulated infra-red emission of trivalent uranium, *Phys. Rev. Lett.* **5**, 557–559 (1960).
2. E. Snitzer, Optical maser action of Nd^{3+} in a barium crown glass, *Phys. Rev. Lett.* **7**, 444–446 (1961).
3. L. R. Solon, Lasers and fusion, *Ind. Res.* **13** (11), 31–34 (1971).
4. CGE gives details of experiment that produced fusion neutrons, *Laser Focus* **5**, Dec. 14 (1969).
5. Hundred-joule lasers are producing high-temperature plasmas, *Phys. Today* **22** (11), 55 (1969).
6. *Lasersphere* **3**, June/July, 9–10 (1973).
7. *Optical Spectra* **6**, Sept., 21–22 (1972).
8. *Optical Spectra* **5**, Apr., 17–18 (1971).
9. M. Eleccion, The family of lasers: A Survey, *IEEE Spectrum* **9**, Mar., 26–40 (1972).
10. C. K. N. Patel, Selective excitation through vibrational energy transfer and optical maser action in N_2–CO_2, *Phys. Rev. Lett.* **13**, 617–619 (1964).
11. *Int. Sci. Tech.* **Jan.**, 61 (1967).
12. D. A. Leonard, *Appl. Phys. Lett.* **7**, 4 (1965).
13. A. E. Hill, *Appl. Phys. Lett.* **12**, 324 (1968).
14. *Microwaves* **10**, Mar., 38 (1971).
15. *Ind. Res.* **15** (6), 23 (1973).
16. A. J. Demaria, Review of C.W. high-power CO_2 lasers, *Proc. IEEE* **61** (6), 731–748 (1973).
17. W. H. Christiansen and A. Hertzberg, Gas dynamic lasers and photon machines, *Proc. IEEE* **61** (8), 1060–1072, (1973).
18. S. Yatsiv, Pulsed CO_2 gas dynamic laser, *Appl. Phys. Lett.* **19**, 65–68 (1971).
19. *Microwaves* **10**, Mar., 16 (1971).
20. *Microwaves* **10**, Mar., 18 (1971).

21. *Laser Focus* 9, Sept., 24 (1973).
22. *Phys. Today* 25 (4), 19–20 (1972).
23. *Ind. Res.* 14 (5), 48 (1972).
24. K. G. Hernquist, Noblest of the metal-vapor lasers, *Laser Focus* 9, Sept., 39–49 (1973).
25. *Bell Lab. Rec.* 48, Dec., 343 (1970).
26. H. Kressel, H. F. Lockwood, and M. Ettenberg, Progress in laser diodes, *IEEE Spectrum* 10, May, 59–64 (1973).
27. *Microwaves* 12, Mar., 12 (1973).
28. *Ind. Res.* 12 (11), 33 (1970).
29. B. B. Snavely, *Yale Sci.* 47, Dec., 9 (1972).
30. *Lasersphere* 3, Aug. 15, 20 (1973).

8

LASERS IN MEASUREMENT

Extensive use is being made of the straight-line property of a narrow laser beam for measuring *alignment* in various applications. The laser's high frequency coherence has made it useful in applications for high-precision measurements of length, strain, velocities, distance, and even the size of extremely small particulate matter in air or other gases. Other measurement applications include the monitoring of pollutants in the air and noncontact measurements of various products.

8.1. Alignment

Lasers are finding uses in a broad range of construction activities.[1] Providing means of precise alignment, as shown in Figure 8.1, lasers are used in work on tunnels, mine shafts, dredging, bridges, building foundations, brickwork, and sewer and water lines. In laying pipe, such as clay sewer pipe, the light beam is aimed at a reflecting target; this eliminates much of the preliminary stake-out work. The use of lasers during installation below ground obviates the need for repeated transit and level work to assure proper grade. In one instance, the use of a laser permitted the installation of more pipe and required one less man in a six-man crew.

The U.S. government, in a 15-year program, has been using lasers for determining much more accurately the positions of the approximately one million bronze markers which constitute the National Horizontal Network.

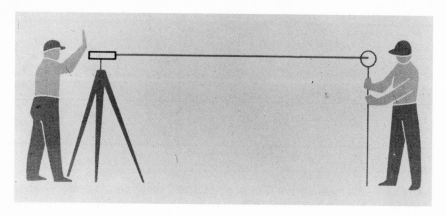

FIGURE 8.1. The sharp straight-line beam of the laser is useful in surveying and other alignment applications.

A six-months section of this program, covering sites in Michigan, Minnesota, and Wisconsin, began in June 1974.

Investigations have revealed that heavily damaged cars often return to the road with their chassis out of alignment. Conventional methods for determining whether or not they are out of line, however, are time-consuming and costly, as well as unreliable. A laser device for the fast and reliable measurement of a buckled car chassis during and after repair work has been developed.[2] The device is comparable to a giant sliding caliper whose jaws have been replaced with a movable laser beam. When the beam is aligned with given points on a chassis it can measure the degree to which the frame departs from complete symmetry to within 1 mm and without necessitating the removal of car parts. It employs an elevated rail system which runs along one side of the car, on which is mounted the laser unit. The laser is aimed along the rail at a sliding deflection prism to produce two beams, each at right angles to the other. One continues along the rail while the other beam can be aimed at one of the reference points on the chassis. To obviate the need to remove wheels and other car parts, graduated rules with a vertical millimeter scale are suspended from the alignment holes, with bolt heads normally used as reference points. When the prism unit is moved along the rail, a retractable metal tape measure is unrolled and the prism's position is read off the scale. The readings are entered on a form and compared with the dimensions specified by the car manufacturer.

Position-sensing detectors have been mounted on high-rise buildings to monitor movement during periods of high winds. Lasers mounted at several ground points and directed at the detectors on the building permit a determination, with the sensors, of any movements of the building. As the buildings sway in the wind, recorders connected to the detectors monitor building movements to better than 0.001 in. By placing several detectors on the buildings, the twisting and bending of the structure are monitored. Such information is valuable to builders and architects in high-wind areas.[3] In California, a He–Ne laser was used to position dredges (see Figure 8.2) in San Francisco Bay during work on a subway tunnel.[4]

Utilizing a new measurement principle based on patterns of laser light diffracted from a gauge attached to material under test, a laser system provides high resolution with no long-term drift or change in calibration.[5] The instrument has been used in the measurement of long-term structural strain, particularly on concrete. The system can also be used to measure displacements in geophysical and civil engineering and other applications. In operation, laser light waves are diffracted by the edges of two arms which are attached to the opposite ends of the material under test. An interference pattern results whose fringe movements typically magnify changes in edge separation by 200–1000 times.

A laser equipment technique which employs He–Ne gas lasers to achieve accuracies which are five to ten times higher and three times as fast as standard alignment telescopes, is now operational. Similar methods for aligning jigs and fixtures for the Boeing 747 have been in use for some time at plants in Seattle and Everett, Washington (Figure 8.3). Lockheed Aircraft Co. has also begun using laser systems for alignment, and it expects to phase lasers into its assembly line for the C-5A and other future programs, replacing the optical telescope by attrition.[6]

One aircraft builder sees the use of laser systems as inevitable because the size of today's (and future) aircraft involves assembly tools up to 200 ft in length, and alignment tolerances of ± 0.01 in.

A patent was issued in 1973 for using a laser for measurements in football

FIGURE 8.2. A laser being used to position dredges in San Francisco Bay.

FIGURE 8.3. A typical laser precision-alignment system, here used on a Boeing 747 wing skin jig (courtesy of the Boeing Co.).

games. In this system a battery-operated laser system measures the distance a football team needs for a first down. The laser unit is in the forward pole marker, directing a visible beam across the field; if any part of the football is in the beam's path, a reflection is observed, showing that a first down was achieved.

8.2. Distance and Length Measurements

The usual radars are echo-location systems which employ *radio*-wavelength electromagnetic waves. The use of lasers in such applications provides certain advantages.

The more power a radar system puts into a pulse, the better equipped it is to track targets, and optical radar systems are similar in this respect. In the movement toward higher power, CO_2 lasers are often used for the optical radar transmitter. One such laser system uses a master oscillator and three amplifier stages to produce 10-μsec pulses 10,000 times a second, with average power reaching more than a kilowatt and individual pulses as much as 10 kW.[7] This system is also a Doppler system, using the frequency changes in the returned signal to measure target velocity. For this use, the frequency of the laser was controlled to within a few kilohertz since any frequency drift would cause an error in a target's measured velocity. Also, the laser was operated in the pulsed mode rather than in the more usual (for CO_2 lasers), CW mode. In this system, a rotating disk with holes in it was used for the modulator. The pulses generated were fed to a power amplifier tube.

Several laser radars have also been developed in Europe.[8] Ferranti,

C. G. E., and L. M. Ericsson have developed laser range finders for aircraft use, and Siemens has developed a laser distance meter. Also in the U.S., a laser radar technique is being considered (by RCA) for automatic landing systems for aircraft, and lasers have been used as precise distance meters in attempting to measure movement of the earth's crust (for example at the earthquake-prone area called the San Andreas Fault). At Boulder, Colorado a laser earth-strain gauge was used in an attempt to detect gravity waves.[9]

The detail possible in a (scanning) radar record when a *laser* is employed as the radar transmitter is indicated in Figure 8.4. This shows a laser radar record of the airport control tower; it is stated that this picture is generated even when fog, mist, or darkness prevent the tower from being seen visually.

As we noted, laser radars have been used in various aircraft applications. The narrowness of their beams, as demonstrated in Figure 8.4, has made such radars of interest for terrain-following applications, and the U.S. Army Electronics Command recently constructed such a radar-employing laser *diodes* (light heterojunction diodes) for transmitting, using at the receiver end a silicon avalanche detector.[10] The sharpness of the laser beam has also been found useful in an intrusion alarm device (also using laser diodes). The high intensity (due to the sharp beam) allows the transmitter and receiver to be spaced much farther apart than when incoherent light-emitting diodes (LED's) are used; also, because of the sharp beam, they generate fewer false alarms than do comparable microwave or ultrasonic intrusion alarm devices.[10]

The silicon avalanche detector has also been used in a " mini-rangefinder" laser radar developed by the U.S. Army Electronics Command's Combat Surveillance and Target Acquisition Laboratory.[11] The laser radar, which is effective over distances of 200–600 m, weighs only 4.75 lb. The transmitter is a YAG laser utilizing a Q-spoiler made of lithium niobate.

As early as January 1970, at a laser technology conference, a session on laser rangefinders included a description of laser radar developments under-way at five different U.S. laboratories.[12] More recently the U.S. Air Force contracted for a half-million-dollar research and development program for a laser radar system for precise tracking of low-flying aircraft, using an infrared laser. In 1974, the U.S. Army unveiled a laser tracker known as PATS (Precision Automatic Tracking System). It is a mobile system and is to be used initially at the Yuma proving ground for measuring trajectory parameters of aircraft and other vehicles, which, when carrying a reflector, can be tracked at distances up to 15 miles.[13]

Another important military use of lasers involves the illumination of a target from an aircraft and an air-dropped bomb which contains an optical seeker. As the bomb descends, it "homes" in on the laser-illuminated target, permitting the aircraft with its laser illuminator to remain at a safe distance. These devices are referred to as laser-guided bombs.

FIGURE 8.4. A laser radar picture (left) of an airport control tower (right) "painted" from 14 sweeps of the laser radar beam (courtesy of United Aircraft Research Laboratories).

A laser radar altimeter was designed for use on the Apollo 15 lunar craft for mapping the surface of the moon. It was carried on the command module and used a pulsed ruby laser.[14] Laser range-measurement techniques had earlier been used for measuring the distance from the earth to the moon after the first Apollo crew had landed on the moon (Apollo 11, July 1969, astronauts Neil Armstrong and Buz Aldrin). The astronauts placed 100 fused-silica retroreflectors on the moon's surface to permit laser light pulses from the earth to be reflected directly back to the earth,[15] and the *time* involved in the two-way transit provided the distance measurement (the velocity of light in free space being known). Those first measurements were accurate to within 30 cm; in 1974, it was announced that more recent measurements have improved the accuracy to 6 in.[16] Slated for launch in 1976, NASA's laser geodynamic satellite can measure, from its orbit, the relative locations of ground laser-tracking stations.

The use of lasers for determining very *short* distances accurately has also been investigated. At the Massachusetts Institute of Technology a ruling engine for making ruled optical gratings was made more precise by having a very stable laser provide error-correction information.[17] Lasers have also been employed to determine the size distribution of extremely small particulate matter in the air.[18] Developed at the U.S. National Bureau of Standards as part of its "Measures for Air Quality" program, the apparatus causes the air which is carrying the particles to be measured to be blown through the light beam of a CW laser. Light scattered from individual particles is collected by separate annular rings of fiber optics set at 5° and 10° to the laser axis, and detected by separate photomultipliers. (We shall discuss the subject of fiber optics later.)

Although there is always a background of particulates in the air resulting from winds, volcanic eruptions, and other natural sources, it is largely the particulates sent into the air by man that offend both lung and eye. In heavy industrial areas fallout rates of 200 tons of dust per square mile per month are not uncommon. From a physiological standpoint particles in the 0.1 to 5 μm diameter range are of primary interest. Generally speaking, particles smaller than 0.1 μm are believed to be inhaled and exhaled without major interaction in the lung, and particles larger than 5 μm tend to be filtered by the cilia before reaching the lungs.

8.3. Velocity Measurements

A new value for the speed of light, approximately 100 times more accurate than any previous measurements, was determined recently with the aid of laser technology.[19] The measurement showed that light travels at a

FIGURE 8.5. Dr. Ken Evenson, who directed the research in 1973 which led to a new value for the speed of light.

velocity of 186,282.3960 miles per second, plus or minus 3.6 ft. The previous margin of error was 300 ft (Figure 8.5). The measurements, made at the National Bureau of Standards, used five different lasers. The new value will enable astronomers to measure interplanetary distances with radar and lasers more accurately, and will permit more precise tracking of satellites and space vehicles.

The apprehending of speeders through the use of radar or other types of speed-measuring systems always has proved difficult on winding streets or hilly residential sections. A recently developed speed meter combines the precision of laser measuring techniques with the accuracy of a crystal-controlled digital computer.[20] The system is based on accurate measurements of the time elapsed when a moving object traverses a relatively short distance. In operation the car interrupts two laser beams that are aimed across the road. A device on the opposite side of the highway senses the presence or absence of either of the two beams. When the first beam is broken, a crystal-controlled electronic clock is triggered. When the second beam is interrupted the clock stops. A solid-state electronic computer translates the clocked time into miles per hour and displays the results on an illuminated digital readout. A design feature which permits accurate measurement of the car speed is the close proximity of the two laser beams. The beams are only 45 mm apart, and an auto moving at 60 mph cannot appreciably change its speed over this very short distance. Therefore, the speed of the car during its travel from one beam to the other can be considered constant.

A laser interferometer system for obtaining velocity measurements in a variety of situations has been developed.[21] The technique can be used to measure the velocity histories of either specular or diffusely reflecting objects. Its velocity resolution can be as high as ± 0.003 mm/μsec, so that velocities over 0.3 mm/μsec can be measured to better than 1% accuracy. The system utilizes the fact that single-frequency laser light, upon reflection from a moving object, acquires a Doppler shift which can be measured. In this way, measurement of the velocity of hypersonic gas flows can be made with better than 97% accuracy using a ruby laser. The laser is made to generate a minute spark in the gas flow, and the radiation emitted by the spark (as it travels with the gas) first enters a multislit spatial filter, then is detected with a photomultiplier, and is finally recorded on an oscilloscope. Velocity is determined by measuring with an oscilloscope the time between the pulses through the slits. The laser technique has been used in the measurement of gases consisting of hydrogen, N_2O, and air, and traveling at speeds of 1800–3000 m/sec.

Laser techniques for measuring the velocity (5500 mph) of a high temperature (6000°F) rocket exhaust have been employed at the U.S. Air Force Systems Command's Arnold Engineering Development Center.[22]

The use of a laser in an air-speed indicator for aircraft has been under investigation by NASA.[23] In one model, a highly stable CO_2 laser is focused as far as 20 m ahead of the aircraft, where the air is undisturbed by the bow wave. The backscattered light (reflected largely from aerosols) is detected, and the shift in frequency (Doppler effect) measured. From this shift, the speed of the aircraft is calculated and recorded digitally. The system includes, in addition to the (CO_2) laser, a colinear transmitting and receiving telescope, an infrared window, an infrared detector, and the necessary electronics for tracking the Doppler shift. Tests made in clear air (visibility 30 miles), at speeds of 10–520 mph, and at altitudes up to 10,000 ft, indicate the technique to be a significant improvement over present air-speed indicators which employ Pitot-tube air-pressure sensors. These are affected both by boundary-layer effects and by turbulence created by the sensor itself.

Lasers have been proposed for measuring the instantaneous velocity of a rocket sled,[24] as a substitute for the more expensive microwave techniques now in use. A corner cube reflector mounted on the vehicle reflects the incident laser light, and electronic equipment provides a time history of the velocity, distance traveled, and acceleration and deceleration of the sled.

Equipment enabling high school and college physics instructors to perform a direct experiment for measuring the velocity of light is available from

Metrologic Instruments, Inc. A football field is used for the location of the experiment as it provides a convenient known length and this also simplifies the mathematics involved. Light from a He–Ne laser located at one corner of the field enters a beam splitter, with some of the laser light being fed directly to a photodetector. The other portion of the light traverses the diagonal dimension of the field and is reflected by a mirror back to the main apparatus where it is combined with the original signal. The laser beam is then modulated, with the frequency of the modulating signal being varied until a null or minimum is observed on the oscilloscope. The null indicates that the distant signal is returning to the detector one half wavelength later than the direct signal. For the diagonal length of the standard football field, the required modulation frequency is approximately 625 kHz. Using the expressions $\lambda = c/f$, is where λ is the wavelength, f the frequency, and c the velocity of light, and $L = \lambda/2$, where L is the traversed, double, distance, the light velocity is then calculated. It should be noted that it is the *modulating* frequency which is playing the important part here, and that a similarly modulated *microwave* beam could also have been used.

In the velocity-measurement category mention should also be made of the use of pulsed-laser "front lighting" of high-speed projectiles for better observation of effects occurring at extremely high speeds. Figure 8.6 shows such a front lighting of a projectile traveling at 15,000 mph.

8.4. Measurement of Acceleration

In addition to distance and the first derivative with time of distance, namely velocity, the second derivative, *acceleration* (or rate), can also be measured with a laser. We noted above that the *short-time* acceleration and deceleration of a rocket sled could be determined from short-time velocity measurements. In recent years, the very *long* time measurements of rates of velocity change have proved to be extremely useful in navigation, particularly for aircraft and long-range missiles. We refer particularly to the *inertial* guidance technique which presently uses three mechanical gyroscopes placed in the vehicle, oriented at the three directions of the x, y, and z axes, and all rotating at high speeds. The *guidance* capability of such inertial guidance systems results from the fact that if any change in velocity or direction occurs in the vehicle's motion, the inertia effect (such as occurs when a mass moving at a constant velocity is forcibly slowed down, i.e., *de*celerated) is provided by the gyroscope because of its tendency to *precess*. We recall that when a spinning gyroscope is placed in a horizontal position and supported at one end, the force (acceleration) of gravity causes the entire gyrostructure to

FIGURE 8.6. Low shock waves and a secondary shock wave for a projectile traveling at 15,000 mph as shown by the use of a patterned background with the front-lighted laser system at the U.S. Air Force Systems Command's Arnold Engineering Center.

rotate or precess about its supporting point. In many gyroscopic inertial guidance systems the gyros are fixed (these are called strapdown systems), but their *tendency* to precess still generates forces on their supports, which can then be measured and thus provide the rate-of-change-of-velocity information. This rate-change information, if accurately determined, can be used to calculate the course of the vehicle. Inertial guidance systems are widely employed in aircraft which are traversing areas such as the Atlantic and Pacific Oceans, where other forms of navigational information (such as aircraft beacons, the radar observation of land contours etc.) are unavailable.

One might wonder how a laser could be substituted for the high-speed rotating mechanical gyros in such systems. Up to this point we have shown lasers as having their reflecting mirrors at opposite ends of their active rods or tubes. If, however, we place three mirrors on a large circle and position them at 120° to one another, light waves can continuously traverse this equilateral path in either a clockwise or counterclockwise direction. If we now place a gas laser tube in one of these triangle legs, it can again experience lasing action from waves passing along its center line in both directions, and function, therefore, as a normal laser. In this arrangement (called a ring laser or laser gyro), any rotation of the system about an axis perpendicular to the plane of the laser light paths will provide an up-Doppler shift for one (say, clockwise) light-path direction, and a down-Doppler for the other. This phenomenon was predicted by Michelson in 1904,[25] but a more accurate analysis involves relativity formalism.[26] Stated simply, the rotation, or rate change, of the laser gyro can be obtained by measuring the frequency difference between the clockwise and counterclockwise laser light paths, and this information is exactly that which a mechanical gyroscope system provides.

In 1973, the U.S. Naval Weapons Center placed a 2.5-million-dollar contract to develop a three-axis inertial guidance system employing ring-laser gyros to provide (1) a midcourse guidance (of a missile), (2) an autopilot function, and (3) an interface with the terminal seeker.[27] Holloman Air Force Base recently contracted for a two-axis ring-laser gyro for a rocket sled, to be tested at stresses as high as 100 g (g being the acceleration of gravity).

8.5. Atmospheric Effects

Lasers are capable of measuring the density of fog, and offshore fog banks are being detected for the U.S. Coast Guard at ranges up to 3 miles with a laser system developed at the Transportation Systems Center (Federal Department of Transportation) in Cambridge, Massachusetts (originally the NASA Electronics Research Center). The laser system appears to be more effective than previous systems and is capable of automatically triggering unmanned fog horns when fog banks exist.[28]

The rapid growth of air travel has placed strong emphasis on air safety, with the result that a phenomenon referred to as "clear-air turbulence" has received much attention. This is a turbulence not in or near convective clouds or thunderstorms but which produces sudden and sometimes violent vertical motions of high-speed aircraft. It has been found that lasers can detect these turbulent regions, and a laser-*Doppler* clear-air-turbulence detection system is under development at the NASA–Marshall Space Flight Center in Huntsville, Alabama.[29] The system is basically a laser radar designed to detect the turbulent regions much as an ordinary radar detects solid reflecting objects.

Air turbulence is also generated *behind* large, high-speed aircraft due to the plane's motion. It has been known to persist for some time after a large jet has landed or taken off from an airport runway, so that a small, lightweight plane can be seriously affected by the turbulence if it lands (or takes off) along that runway too soon after the big jet has left it. Use of a pulsed ruby laser aimed at the area, with a telescope and low-noise detector for observing the small reflected signal, has been proposed to reduce this danger to small aircraft.[30]

8.6. Sonic Boom

Another use of lasers has been investigated for reducing or eliminating an atmospheric phenomenon associated with supersonic aircraft and known as the sonic boom. When an aircraft flies at supersonic speeds its leading point generates a sonic bow wave (shock wave), and its rearmost point tends to bring that region of the atmosphere back to its undisturbed state. It has long been known that the loudness of the sonic boom is related to the length of the craft, with very long aircraft generating weaker sonic effects. Several research centers are exploring the possibility of having a powerful laser beam focused perhaps several hundred feet in front of a supersonic aircraft, with this focal point, through its ionizing action, acting as the most forward part of the aircraft and initiating, at that point, the shock wave. It is hoped that this effective added length will reduce the severity of the sonic boom.[31]

8.7. Measurement of Pollutants

Laser radars have been found useful in detecting and measuring air pollution.[32] They have been found to be particularly effective in detecting thin atmospheric clouds of particles which are invisible to conventional radar. The particles detected are at least 1000 times smaller than the minimum which standard microwave radar can detect. At a range of 100 miles, a cloud density of 100 particles per cubic foot, with the particles averaging 20 μm (0.0002 cm) in diameter, produces echoes for laser radars. Both smog particles and atmospheric cloud particles are usually in the 10- to 50-μm range, so that this procedure shows promise for air-pollution measurement.

Combined use of a laser and a computer has permitted quick detection and measurement of concentrations of pollutant gases.[33] Capable of identifying concentrations as low as 1 part in 10 billion, one system uses a technique similar to gas spectroscopy. Specific gases leave their "fingerprints" on laser light at certain wavelengths. The amount of light that is absorbed shows the amount of pollutant in the air. The computer controls the laser, tuning it through the wavelengths of various pollutant gases. Light energy absorbed by a gas increases the temperature and pressure of air in an opto-acoustic absorption cell in proportion to the quantity of gas. A sensitive microphone in the cell detects the increase in pressure and converts it to an electrical signal that is fed to the computer.

Lasers are employed in a method for monitoring nitric oxide (NO) in such source emissions as automobile exhausts and electrical power plant stacks.[34] Nitric oxide is a major air pollutant and is the precursor to nitrogen dioxide (NO_2), the trigger molecule in photochemical smog formation. When high levels of this pollutant are present eye irritation and other discomforts are severe. The monitor's action is based on the Zeeman shift of a NO absorption line into coincidence with a CO laser line. A static magnetic field shifts the NO absorption line into near coincidence with the laser line. Measurement with this apparatus on samples of known NO content indicate that quantities as small as 60 parts per million of NO are detectable. Recently, Bell Laboratories engineers made pollution measurements in the stratosphere, 17 miles above the earth. The primary purpose of the test was to determine the effect of the sun (after sunrise) on the NO concentration in the stratosphere.[35]

Lasers have also been used to detect and measure NO_2.[36] Unlike other pollutants, NO_2 has a broad absorption band and fluoresces when excited by any visible laser light. Air samples containing NO_2 have been excited with a helium–cadmium laser at 4416 Å and an argon ion laser at 4880 Å. A conventional photomultiplier measures fluorescence in the 6300- to 8000-Å range by counting the number of light quanta generated in a present period. The strength of the laser beam is also monitored, since it affects the light output. We shall discuss shortly the possibility of detecting pollutants at a distance by means of the Raman scattering of laser light.

8.8. Lasers in Machine-Tool Measurements

Lasers have been used in a variety of machine-tool applications, (1) as an integral part of machine-tool positioning and control systems for sensing and for correcting machine-tool geometry, and (2) in precision inspection applications.[37] Figure 8.7 shows a laser interferometer gauge. In this device, the beam from the laser (center left) enters the center structure and is there

FIGURE 8.7. A laser probe with a 1.5-inch range for measuring internal and external diameters of bearings and rings.

divided into two beams by a beam-splitter. One beam is reflected off a mirror in the beam-splitter, and the second beam is reflected from a mirror attached to a movable stylus. The two reflected beams are combined to form constructive and destructive interference regions (light-wave fringes) by virtue of the moving mirror. Sensors count the number of fringes observed as a function of the displacement of the probe (the movable stylus), thereby providing a highly accurate length measurement, without the usual requirement of so-called gauge blocks or masters (blocks having precisely established lengths). A typical ID–OD (inner diameter and outer diameter) laser-bearing gauge can measure an ID range of 0.078–0.500 in. and an OD up to 1.156 in.

The original Michelson, moving-mirror, technique, as shown in Figure 8.8, can also be used for fringe counting. In the contacting probe device of

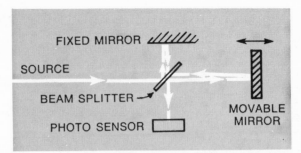

FIGURE 8.8. The optical system of this laser interferometer is similar to the one used by Michelson to measure the length of the standard meter bar.

Figure 8.9, the mirror motion is effected by a carbide-tip ball at the far left of the figure; it can be seen more readily in the close-up photograph of Figure 8.10. This figure also shows the fringes which are generated (by the 45° optical plate of Figure 8.8) at the position of the photosensor of that figure. Movement of the ball-contacting probe causes the fringes to move past the sensor. They can thus be counted and the amount of probe motion determined. The probe is mounted on an air bearing.

Laser systems can also be used as the feedback devices on very accurate machines, including very large, numerically controlled measuring machines. As shown in Figure 8.11, a portable, laser-interferometer length-measuring system can be used to check the feedback devices and slide motions of high-

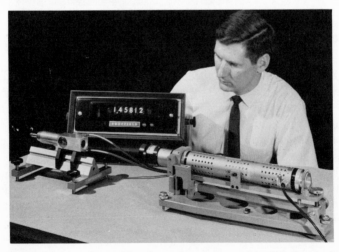

FIGURE 8.9. A contracting probe using a laser interferometer for length measurements.

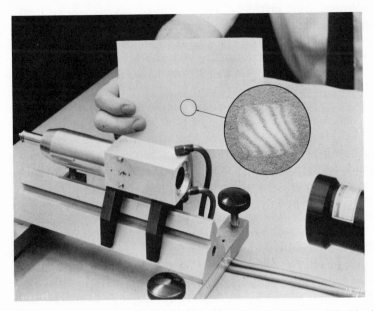

FIGURE 8.10. A close-up view of the measuring system of Figure 8.9. The insert shows the fringes which are counted by the photosensor.

FIGURE 8.11. A portable laser-interferometer system can be used to check the accuracy of feedback devices and slide motions of large, numerically controlled, measuring machines.

precision measuring machines. The devices can themselves be used as the feedback systems on numerically controlled machine tools. The advantages of using laser interferometers in such applications as these are that (1) they require, as we noted, no mastering with gauge blocks, and (2) they make possible machine checkout over lengths for which no tangible physical standards are available.

A noncontact optical system for dimensional measurement which uses a laser light source is shown in Figure 8.12. It operates by projecting a small spot of light from the laser onto the part surface, and it senses the movement of this spot over the surface. The probe produces an output signal whose magnitude and phase indicate the deviation of the distance between the part and the probe from some preset null position, and also the direction of this deviation. This output may be used to drive a servo system moving the slide on which the probe is mounted so as to keep the probe a constant distance from the part surface. The figure shows one arrangement of the equipment. The light is brought to the probe from the laser source at the right through a fiber-optic bundle (the loop being held by the operator), so that the laser itself does not have to be moved. The overall system resolution is better than 0.0005 in., and it may be used to gauge any type of part surface regardless of its reflectance. One of the many applications of this gauging probe is non-

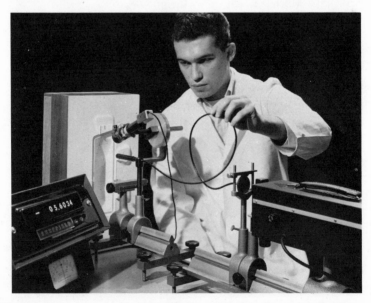

FIGURE 8.12. A noncontacting optical system for dimensional measurement which uses a fiber-optic bundle to bring the light from the laser source to the probe.

contact measurement of easily deformed soft parts such as automotive-body clay-model mock-ups. Such a system significantly reduces the time lag between the clay model and the body dies.

8.9. Holographic Microscopy

Numerous new measurement techniques in the fields of physics and medical research have become available since the advent of the laser and the hologram concept.

Holographic microscopes are now offered as standard laboratory systems [38] where the third-dimension property of holograms permits, in a single exposure, the equivalent of over a hundred exposures with an ordinary photographic microscope. They thus permit detailed analysis and measurement of fibrous subjects such as nerve cells and plastic fibers through the procedure of focusing in the third dimension. Double-exposure holography is also usually possible on the commercial-microscope models, permitting the use of the process of holographic interferometry (a subject we shall discuss shortly). Figure 8.13 shows how the optical-laser interference fringes provide a contour mapping of the object under examination. The completely undistorted three-dimensional viewing and recording of important events (such as the way in which human blood cells devour bacteria, or the way in which living cells divide) is made possible by holographic microscopy. [39]

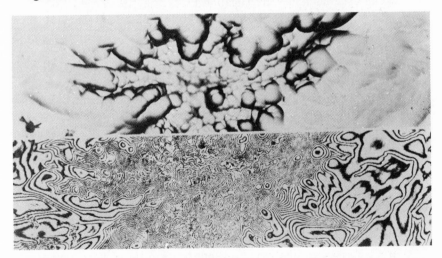

FIGURE 8.13. Conventional bright-field microscopy yields view (top) of a glass slide etched with hydrofluoric acid. The same slide is contour-mapped in a holographic interference microscope (courtesy of Mary E. Cox, University of Michigan).

Numerous optical arrangements have been described for holographic microscopes, including geometric, corrected, and uncorrected optics, and it has also been noted that hologram speckle (a laser characteristic we shall discuss shortly) can be troublesome in this form of microscopy.[40]

8.10. Saturation Spectroscopy

The science of spectroscopy also benefited substantially from the advent of lasers. The narrowness of the frequency line of the light from a laser has enabled scientists and engineers to exploit a new class of spectroscopic techniques, often referred to as saturation spectroscopy.[41] The exact, single frequency necessary to excite an atom to a particular energy level can, by *theory*, be accurately expressed, but because of the random movement of atoms, a Doppler *spread* of frequencies has always been observed. We noted earlier that the theoretically expected narrow-band *radiation* of atomic transitions is also not observed because of the Doppler shifts in radiated light from the randomly moving radiators. Only when the radiators are held rigidly fixed, as by Mössbauer's technique, do these radiation lines approach their expected, narrow-band values. For many spectroscopic measurements, the Doppler spread in excitation is a limiting factor.

A laser beam moving in one direction can saturate one of the absorption frequencies of a gas under investigation by exciting those atoms which are moving at right angles to the beam and therefore possess no Doppler effect. A second beam from the same laser, caused by mirrors to be moving in the exactly opposite direction, will experience a completely different situation; for it, all of the 90°, zero-Doppler atoms will have already been excited by the oppositely moving saturation beam. This second beam will, therefore, experience practically no attenuation in passing through the sample. It was in this way that the Lamb shift (after Willis Lamb, a U.S. Nobel laureate), which is the separation of two closely spaced energy levels in an atom having one electron, was first measured directly, in hydrogen, at visible-light frequencies.[42]

8.11. Raman and Brillouin Spectroscopy

Sir C. V. Raman discovered an interesting form of light scattering in 1928; for this discovery he received the Nobel Prize in Physics in 1930. Since then his name has been associated with this *inelastic photon scattering*. When powerful light, such as that emitted by a laser, strikes a sample, most of the photons are scattered elastically, but a few undergo inelastic scattering. This produces a spectrum of frequencies in the scattered beam which can characterize and therefore identify the scattering molecule. Those photons which lose energy appear on the lower-frequency side of the exciting light, and those that gain energy appear at higher frequencies. The primary energy exchange between the exciting photons and the molecule is between the vibrational energy levels of the molecule, so that the scattered photons have shifts in frequency which are highly characteristic of the vibrational energies of the molecule. Theory can indicate that a molecule should have a particular vibrational pattern and, therefore, a corresponding Raman spectrum.[43]

In the past, limitations on the sensitivity of the analyzing instruments have prevented observation of some of these spectral lines. Also, ordinary light sources, even very powerful arc lights (e.g., the Toronto arc), often require hours of exposure to reproduce, on film, spectral records which can be completed in seconds using a laser as the source.[44] Weak Raman lines that had not previously been observed have now been seen with a laser. Also, lasers have produced stimulated emission, causing some strong Raman lines to actually lase. We mentioned earlier that this phenomenon is referred to as stimulated Raman scattering or stimulated Raman emission. Laser–Raman spectroscopy equipment is now available commercially.[45] Such systems can resolve two lines spaced only 0.08 Å apart at the He–Ne laser wavelength range (6328 Å).

Because certain gases which are responsible for air pollution (SO_2, CO_2, and N_2O_5, in particular) have well separated Raman lines, methods of using a laser for monitoring industrial stack gases at a distance are possible.[46]

A second form of inelastic scattering of light is caused by the periodic fluctuations of refractive index, caused by thermally excited sound waves (phonons) in the scattering medium. Single-frequency (laser) light scattered by such substances again acquires a spectrum comprising frequencies higher and lower than the exciting frequency. These are called Brillouin or Doppler components and, as in the case of Raman scattering, the lower-frequency component (lower side band) is referred to as a Stoke's line and the higher as an anti-Stoke's line. The use of lasers to excite the Brillouin scattering was described as early as 1964.[47,48]

8.12. Mass Spectroscopy

A device known as a time-of-flight mass spectrometer has been quite useful in determining the constituents of materials. The material is first ionized and the ions subjected to a dc electrostatic field. The ions move along the direction of this field over a moderately long path, and their time of arrival at a plane is detected electronically. High-mass ions acquire less acceleration than low-mass ions, and hence the arrival time gives an indication of the masses of the various ions present in the material under study. The laser is advantageous in such time-of-flight spectrometers because it provides a precisely timed instant for the ionizing process. Focused laser pulses are made to vaporize the sample, thereby producing the ions for the spectrometer.[44]

8.13. Holographic Interferometry

If two successive holograms of the same object are superimposed in the same hologram emulsion, an image of the object covered by interference fringes may be reconstructed. These fringes are a direct measure of the

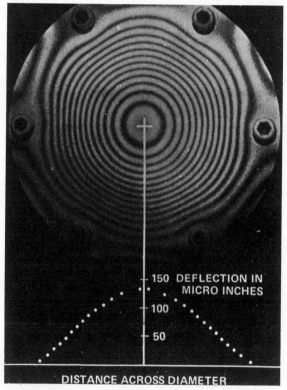

FIGURE 8.14. Holographic interferometry permits precise measurement of a 1-mm-thick diaphragm displacement when it is subjected to a small gas pressure (courtesy of Canadian Research and Development).

topographical changes of the object which occurred between the two exposures.[49,50] The principle of this type of holographic interferometry is readily explained. Two image-forming wave-front components are reconstructed from the hologram, and any slight difference between the two wave fronts will manifest itself in the form of an interference pattern, which may be photographed. Holographic interferograms permit study of deformations in objects of great complexity; moreover, the objects need not be mirrorlike, as had been required in earlier optical-measurement techiques, but may, in fact, be perfectly diffusing. Figure 8.14 shows the effect of a slight pressure applied to an aluminum plate. The first holographic exposure was made with the diaphragm subjected to a slight gas pressure from behind. The second exposure was made with the pressure relieved. The relative displacements of the diaphragm can be derived from the fringe map, as shown in the graph beneath the reconstructed double hologram. This procedure has also been used in medical applications, in which the two (pulsed) laser illuminations (50 nsec) occur in rapid succession (150-μsec intervals). Such double-pulsed ruby-laser interferograms have shown the expansion of a person's chest and the movement of the abdomen during breathing, and appear useful in similar experiments in biomechanics, neurology, and speech research,[51] Figure 8.15 shows a portable interferometric holocamera which is able to show such body movements. One model permits a (minimum) pulse width of 3×10^{-9} sec to be achieved, and a minimum pulse separation (between the double pulses) of 200×10^{-9} sec. Figure 8.16 portrays the fringes generated by even slight motions of a person which occur between two closely spaced pulses.

8.14. Interferometric Structure Analysis

The interferometric procedure is also useful for analyzing vibrating structures, such as loud speakers and sonar transducers. In such reconstructed holograms, the fringes represent lines of equal displacement during the time between the two laser flashes. The technique can be used to analyze vibrational modes, and to detect hidden flaws which produce localized, nonuniform vibrations.

The analysis of the heating and the accompanying expansion of solid-state components is also accomplished readily with interferometric holography. Thus, two exposures of a resistor, before and after current is applied to it, provide a fringe photograph showing the expansion of the resistor and the surrounding supporting material (e.g., epoxy).

The measurement of the bending properties of plates has been demonstrated with interferometry.[52] It has been known that if an accurate measure

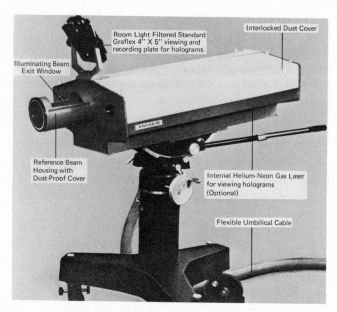

FIGURE 8.15. A portable double-pulse ruby laser for interferometric applications (courtesy of Hadron, Inc.).

can be made of the displacement of a rectangular bar which is subjected to pure bending, Poisson's ratio for the bar can be established. Figure 8.17 is a diagram of the test arrangement used for such a determination. When the test plate is subject to bending, it is known that the contour lines are hyperbolas, and from these hyperbolas, Poisson's ratio can be determined. Figure 8.18 shows a sample reconstructed interferometric hologram. The sharp hyperbolic contours are quite prominent.

8.15. Interferometric Detection of Footprints

Most composite or fibrous materials, such as wood, plastics, and woven material, exhibit the phenomenon of creep after being deformed. Thus, a footprint deforms, often in very minute ways, many floor surfaces, including rugs or carpets. When the foot is removed, the fibers of the material do not immediately return to their original shape, but creep back slowly, sometimes taking hours to do so. Such movements, however, even with velocities of only a few nanometers per second, are detectable by laser holographic interferometry. A hologram is first made of the area where the impressions are suspected and some minutes later a second hologram is exposed on the same

FIGURE 8.16. Reconstruction of a double-pulse (interferometric) hologram of a person. Fringes are most clearly seen at the lower left portion of the figure.

photographic plate. Any minute differences in the reflected light resulting from movements in the suspected area appear as interference fringes.[53] Figure 8.19 shows a typical reconstructed hologram made by this technique.

8.16. Interferometric Testing of Rocket Components

The holographic interferometry procedure has found wide use by U.S. space scientists for seeking out possible flaws in materials for the rocket engine

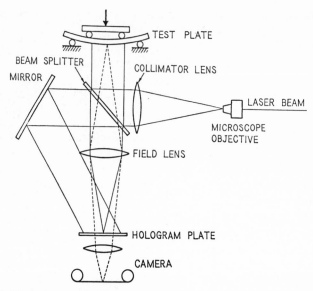

FIGURE 8.17. An experimental arrangement used for determining Poisson's ratio through holographic interferometry.

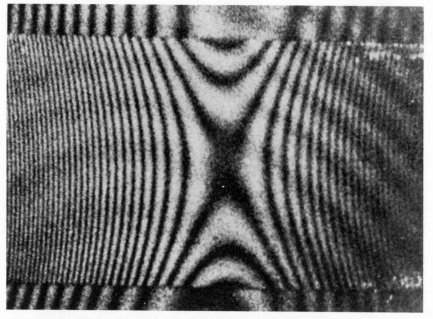

FIGURE 8.18. An interferogram for a steel plate 30 mm in width.

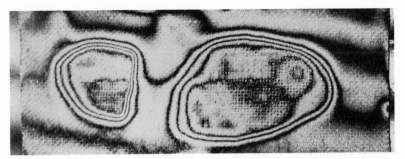

FIGURE 8.19. The outline of a footprint on a carpet, although not visible to the naked eye, is revealed in a reconstructed laser interference hologram.

being developed to power the Space Shuttle.[54] In this new method of non-destructive testing, the equipment is designed to be portable so as to be usable at remote sites outside a laboratory.

8.17. Nonoptical Holographic Interferometry

The use of the double-hologram technique has been investigated at microwave frequencies; the longer wavelengths permit much larger structures to be examined (because of the longer coherence length). Papi *et al.*[55] used the scanner of Figure 8.20 to produce the microwave fringes of Figure 8.21; these resulted when a second exposure of the deformed test object was superimposed on the first (Figure 8.22). The use of microwave interferometry in synthetic-aperture microwave systems[56] has been proposed for examining large structures (Figure 8.23). We shall describe the synthetic-aperture technique in a later chapter. Such techniques may also be useful in acoustic holographic systems.[57]

8.18. Laser Speckle

The property of laser-illuminated objects and surfaces which generates glints or a sparkling effect is called "speckle." Almost all directional patterns of radiating devices possess, in addition to the main radiation beam (the main lobe), so-called side lobes, with phase reversals occurring at successive lobes. The reflection pattern of a very flat reflecting surface when illuminated with coherent wave energy also exhibits many side lobes, also having successive phase reversals. When *many* adjacent, flat, reflecting areas, all oriented slightly

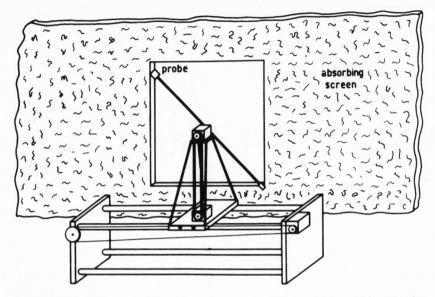

FIGURE 8.20. The microwave scanning arrangement used in making the first microwave holographic interferogram. The signals picked up by the scanning probe were recorded for the unstressed test object, and the signals picked up during a *second* scan, with the test object stressed, were superimposed upon the first set.

FIGURE 8.21. Microwave fringes (the roughly circular shapes at the right) show the extent of the change in the test object.

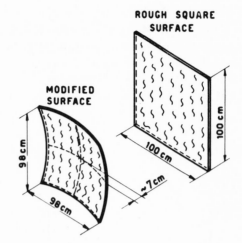

FIGURE 8.22. A sketch of the test objects used in the two previous figures. A square reflector was modified as shown to generate the microwave fringe pattern.

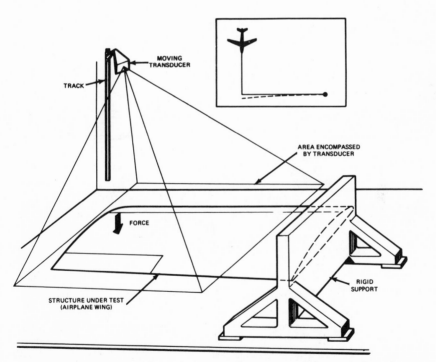

FIGURE 8.23. The use of the holographic synthetic-aperture technique for generating a microwave holographic interferogram. The long coherence length of microwave signals permits the accurate measurement of the deflection of extremely large structures such as aircraft wings by this interferometric procedure. The records obtained will be similar to air-borne synthetic-aperture radar records of a wall (upper insert).

differently, are illuminated with coherent wave energy, each area generates its own reflection beam pattern, and some of its negative minor lobes destructively interfere with positive lobes of other flat areas. If the coherent illumination remains constant, a detector, moved about so as to measure the reflected pattern from the stationary object, would experience very strong signals (constructive interference) in certain directions, and very weak signals in

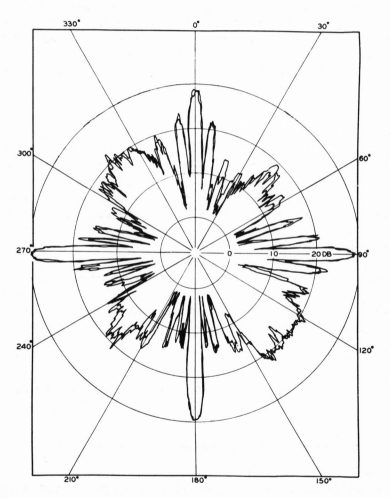

FIGURE 8.24. A typical acoustic reflection pattern involving single-frequency sound waves shows large peaks and dips, generated by phase additions and cancellations and similar to the "speckle" pattern generated by the single-frequency light from a laser (from Reference 57).

others. An example of this is shown in Figure 8.24 for the case of an object irradiated with coherent sound waves.

This is exactly what happens when a highly coherent light source, such as a laser, illuminates almost any normal object. The wavelengths of light are so short that even extremely tiny areas, which are "flat" and which exist on the object, generate for the highly single-frequency laser light waves, sharp reflected beams, each with its many side lobes. As a viewer moves his head about while observing such a laser-illuminated surface, many bright glints of light are observed which appear and disappear with the head motion. It is this effect which is called "speckle." When the observer is at a large distance from the illuminated surface, he must move his head farther to observe the same effect. Since the beam width θ of the plane reflecting area is related to the distance r and the arc length S by the equation

$$S = r\theta$$

this effect is to be expected. As noted by B. M. Oliver,[58] the speckle effect is a "random, but *stationary* diffraction pattern." It is not observed with ordinary light because the frequency coherence of almost any light source except a laser is such that a large number of frequencies are involved; for each of these major lobes have different widths and the minor lobes are differently spaced. Oliver also noted that laser illumination of milk produces no speckle because milk is a colloidal suspension, and the Brownian motion causes the diffraction patterns to move about very rapidly.

The speckle problem also exists when *holograms* are illuminated (reconstructed) with a laser. Gabor has pointed out that, for optical holograms, the speckle effect can be largely eliminated by employing a reconstructing source which at least *approaches* the incoherence of ordinary white light. He recommends the use of one of the lines of the mercury arc light as a source, a procedure also found effective in 1965 by one of the author's associates, L. Rosen. Thus, in making Figure 4.8, one line of a mercury arc line (the green) was used as a reconstructing source. Even when visually observed, this reconstruction with the mercury arc green line showed unnoticeable speckle.

8.19. Measurement Techniques Employing Speckle

It had been noted fairly early that the speckle pattern of a vibrating object changes with the period of vibration. If the speckle pattern is observed on a TV monitor, the frame rate of the TV system plays a part in the appearance of the viewed pattern. If the period of vibration is longer than one frame time, averaging of the signal takes place in the vidicon pickup tube. As a result, the speckle pattern will vanish at all points at which the extent of the motion is comparable to a fraction of a wavelength, and will remain unchanged along the nodal lines.

Certain advantages are gained when *interferometric* speckle hologram patterns are made; most of the research in this field has been conducted at the British National Physical Laboratory and at the Loughborough University of Technology.[59]

8.20. References

1. *Nation's Bus.* **Jan.**, 60 (1968).
2. *Ind. Res.* **14** (9), 21 (1972).
3. *Opt. Spectra* **7**, Apr., 42 (1973).
4. *Microwaves* **9**, Sept., 54 (1970).
5. *Lasersphere* **3**, Aug. 15, 22 (1973).
6. *Aerosp. Technol.* **Apr. 8**, 24 (1968).
7. *Electronics* **4**, May 13, 40 (1968).
8. *Laser Focus* **9**, Aug. (1973).
9. *Phys. Today* **24** (6), 19 (1971).
10. *IEEE Spectrum* **10**, May, 61 (1973).
11. *Lasersphere* **2**, Jan. 15, 9 (1972).
12. Fourth DOD Conference on Laser Technology, Jan. 6–8, 1970, U. S. Naval Training Center, San Diego, Calif. The author presented an invited paper there on recent developments in holography at the conference's opening session.
13. *Opt. Spectra* **8**, Jan., 26 (1974).
14. *Lasersphere* **1**, Aug. 23, 1 (1971).
15. A. L. Hammond, Laser ranging: Measuring the moon's distance, *Science* **170**, 1289–1290 (1970).
16. *Opt. Spectra* **8**, Jan., 18 (1974).
17. Laser applications section, *Ind. Res.* **12** (3), L7 (1970).
18. *Opt. Spectra* **7**, Jan., 35–36 (1973).
19. *Opt. Spectra* **7**, Jan., 41 (1973); *Ind. Res.* **16** (7), 15–16 (1974).
20. *Ind. Res.* **13** (5), (1971).
21. Interferometric velocity, *Ind. Res.* **15** (4), 8 (1973).
22. *Lasersphere* **1**, Apr. 5, 1 (1971).
23. *Lasersphere* **3**, Jan. 15, 8 (1973).
24. *Opt. Spectra* **7**, Apr., 42–43 (1973).
25. A. A. Michelson, Relative motion of Earth and aether, *Phil. Mag.* **8**, 716 (1904).
26. W. A. Shapiro, Ring lasers, *Bendix Tech. J.* **Summer**, 48–57 (1969).
27. *Lasersphere* **3**, Aug., 22 (1973).
28. *Opt. Spectra* **5**, Apr., 9 (1971).
29. *Opt. Spectra* **5**, Dec., 33 (1972); *Lasersphere* **3**, Sept. 15, 10 (1973).
30. *Opt. Spectra* **7**, Apr., 42 (1973).
31. B. R. Eichenbaum, Methodology for sonic boom abatement employing high-power lasers, 1969 Hadron Corp. Technical Report 5246.
32. *Electron. News* **14**, Aug. 11, 16 (1969).

33. *Opt. Spectra* 7, Mar., (1973).
34. *Opt. Spectra* 7, Nov., 38 (1973).
35. *Opt. Spectra* 8, June, 16 (1974); *Bell Lab. Rec.* 52, Oct., 298–299 (1974).
36. *Electronics* 45, Feb. 14, 32 (1972).
37. W. E. Kock, The use of lasers in machine tools, *Opt. Spectra* 4 (March 1970), presented at the International Laser Colloquium, Paris, France, November 1969. Also, in French; W. E. Kock, "L'emploi des lasers dans les equipements de mesure," *Mesures* 35, 75–80 (1970).
38. *Laser Focus* 7, Apr., 35 (1971).
39. *Ind. Res.* 14 (1), L1 (1972).
40. M. E. Cox, Close-up of holographic microscopy, *Laser Focus* 7, Feb. 41–43 (1971).
41. W. D. Metz, Physics with lasers: High resolution coming of Age, *Science* 175, 739–40 (1972).
42. T. W. Haench, J. S. Shahin, and A. L. Schawlow, *Nature (London) Phys. Sci.* 235, 63 (1972).
43. J. L. Koenig, Laser Raman spectroscopy, *Res. Dev.* 20 (6), 18–22 (1969).
44. T. H. Maiman, Laser applications, *Phys. Today* 20 (7), 24–28 (1967).
45. *Laser Rev.* Mar., 3 (1969).
46. M. C. Johnson, Laser–Raman remote gas sensing, *Bendix Tech. J.* Spring, 55 (1971).
47. R. Y. Chiao and B. P. Stoicheff, Brillouin scattering in liquids excited by the He–Ne maser, *Appl. Opt.* 3 (10), 1286–1287 (1964).
48. G. B. Benedek, J. B. Lastovka, K. Fritsch, and T. Grevtak, Brillouin scattering in liquids and solids using low-power lasers, *Appl. Opt.* 3 (10), 1284–1285 (1964).
49. R. A. Powell and K. A. Stetson, *J. Opt. Soc. Am.* 55, 1594 (1965).
50. B. P. Hildebrand and K. A. Haines, Interferometric measurements using the wavefront reconstruction technique, *Appl. Opt.* 5, 172 (1966).
51. *Holosphere* 2, Jan., 1–4 (1973), (discussion of a report by G. Norgate and J. B. Adolph which appeared in *Canadian Research and Development*, Maclean-Hunter Ltd., Toronto, Douglas Dingledein, ed.).
52. M. Saito, I. Yamaguchi, and T. Nakajima, Application of holographic interferometry to mechanical experiments, U. S.–Japan Seminar on Holography, Washington, D. C., October 1969. The papers of this seminar are published in *Applications of Holography*, Plenum Press, New York, 1971.
53. *Holosphere* 2, June, 8 (1973).
54. *Holosphere* 3, Apr., 2–3 (1974).
55. G. Papi, V. Russo, and S. Sottini, *Proc. IEEE* 60, 1004–1005 (1972).
56. W. E. Kock, *Proc. IEEE* 61, 135–137 (1973).
57. W. E. Kock, Acoustic holography, in: *Physical Acoustics*, Academic Press, 1973, Vol. 10, W. P. Mason and R. N. Thurston, eds.
58. B. M. Oliver, Sparkling spots and random diffraction, *Proc. IEEE* 54, 220–221 (1963).
59. Speckle at Loughborough, in: *European Scientific Notes*, U.S. Office of Naval Research, London, Oct. 1973, pp. 274–275.

9

MICROWAVE HOLOGRAPHY

If one defines a microwave hologram as the photographically recorded inter-ference pattern between a set of coherent microwaves of interest and a coherent reference wave generated by the same source, then one can say that the first microwave hologram was made at the Bell Telephone Laboratories in 1950.[1] In that work, only the holograms (the interference patterns) themselves were of interest, and no wave reconstruction was performed; nevertheless, the method employed in recording the interference patterns is of interest because it is still in use today.

9.1. Early Developments

In the early experiments, the interference patterns were made visible by photographic scanning. In making the pattern, two wave sets from a common source of coherent microwaves were used, as shown in Figure 2.6. The waves of interest in this case were those passing through the metallic microwave lens on the left; the reference beam issued from the microwave horn shown at the far right. A small probe antenna, with a small neon lamp attached, scanned the field. The probe signal was amplified and fed to the neon lamp, causing its brightness to vary. A camera set at time exposure thus recorded the otherwise invisible microwave interference pattern, (a microwave holo-gram) as shown in Figure 2.5 and 2.7.

Full hologram concepts were first investigated in the microwave field by J. R. Patty of the U.S. Navy Ordnance Laboratory.[2] Patty also used a scanning probe to sample the interference field created by combining a reference microwave field with the microwaves emanating from back-lighted and front-lighted objects. However, he caused the scanning probe output to intensity-modulate a cathode-ray spot which moved in synchronism with the scanning microwave probe. The cathode-ray tube was then photographed at time exposure, and the photographic record reduced to about ⅛ in. on a side. Patty then illuminated this reduced photograph with narrow-band light (his experiments preceded development of the laser), thereby reconstructing a visual image of the original object as illuminated by microwaves.

More recent experiments along these lines, but with laser readout were reported by Dooley.[3] Figure 9.6 shows one of Dooley's objects, a large, metallic letter *A*. The hologram he obtained is shown in Figure 9.2, and the laser reconstruction is shown in Figure 9.3. Other microwave holography experiments have been reported by Duffy[4] and by Tricoles and Rope.[5] G. A Deschamps[6] has reported on "radio holograms," noting that the reference wave could be introduced electronically so as to stimulate a plane wave, a

FIGURE 9.1. Object for an X-band hologram is a large, metallic letter *A*.

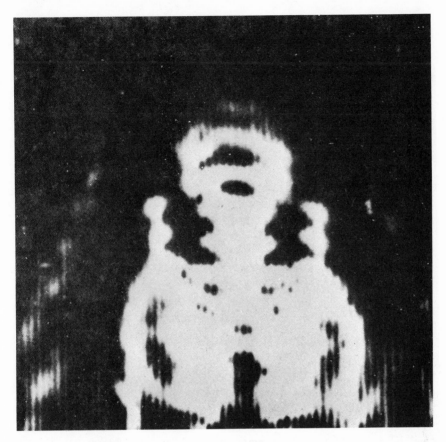

FIGURE 9.2. X-band hologram of the *A* in Figure 9.1.

spherical wave, or, through special phase-shifting procedures, a "slow" wave.

Some limitations of microwave holography are obvious from Figure 2.7 and 9.2. In both figures, X-band (10 GHz) frequencies were used, and it is seen that there are only a limited number of fringes, as compared to those in the usual laser hologram. Accordingly, the detail in the reconstructed image is very limited, as is seen in Figure 9.3. Some improvement is possible by using the shorter-millimeter waves, but a basic limitation still exists. However, because radio waves can penetrate many substances which are opaque to light, e.g., cloth, wood, and stone, some special uses of microwave holograms have become significant. We shall discuss in the next section their ability to indicate the presence of hidden metallic objects such as concealed weapons.

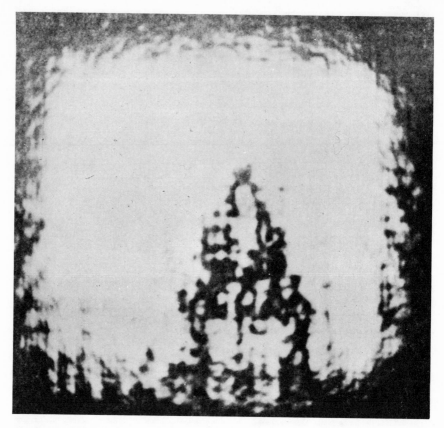

FIGURE 9.3. Reconstructed image from the hologram of Figure 9.2. Resolution is poor compared to laser hologram because of the longer wavelengths of X-band radiation.

9.2. Detection Devices

The ability of microwaves to penetrate materials which are opaque to light waves has been exploited as a means for detecting concealed weapons hidden on persons about to board an airliner;[7] Figure 9.4 is a hologram of a metallic toy gun, and Figure 9.5 is the reconstructed image of the gun, with the barrel pointed toward the right. In these figures, "phase-only" holograms were employed, so that Figure 9.4 is referred to as a "phasigram." The microwave frequency employed was 70 GHz (λ = 4.3 mm) and a spiral motion of the scanning horn was used (as contrasted to the up-and-down

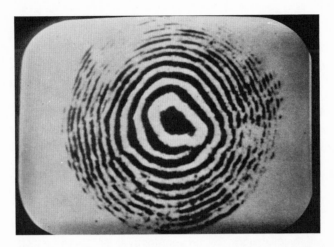

FIGURE 9.4. A microwave hologram of a concealed toy metallic revolver.

motion of Figure 2.6). The spiral scan pattern is evident in Figure 9.4. When the gun was placed in a pocket of a tweed coat or concealed in a synthetic-leather handbag, reconstruction of the corresponding phase-only holograms still gave sufficiently recognizable images of the gun "to allow easy identification." [7]

One annoying detail in this procedure is the time taken to develop and reconstruct the hologram. Recently, an electronic system was developed by Mueller and Goetz of the Bendix Research Laboratories [8,9] in which the

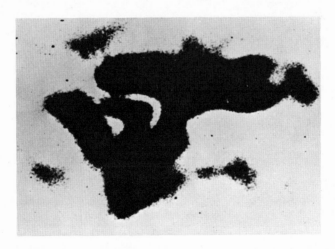

FIGURE 9.5. Reconstructed image of the hologram of Figure 9.4.

microwave (or ultrasonic) holograms can be recorded and immediately re-constructed in real time. An experimental version of this system is shown in Figure 9.6. The equipment consists of a thick deuterium–KDP (DKDP) crystal, an off-axis scanning electron gun, and associated optics (Figure 9.7). In operation, the scanning electron beam is modulated with the holographic information, so that a hologram is written on the crystal in the form of a positive charge pattern. The electric field within the DKDP varies over the crystal according to the holographic signal, thus modulating its refractive index by the electro-optic effect. Coherent light transmitted through the DKDP crystal becomes modulated with (reconstructs) the holographic in-formation. The hologram is periodically erased by flooding the crystal with electrons with an appropriate potential on a nearby grid. The real-time capabilities are presently of the order of 16 frames/sec.

When the hologram (the phasigram) of Figure 9.4 (the toy gun) was sampled and reconstructed by this equipment, photographs very similar to the reconstruction of Figure 9.5 were obtained. Since those phasigrams[7] were on-axis holograms, this constituted a valuable test, as it demonstrated the extent to which the modulator could suppress the dc component. We shall discuss this subject further in Chapter 10.

FIGURE 9.6. This electronic tube has a DKDP crystal which permits real-time reconstruction of both microwave and acoustic holograms.

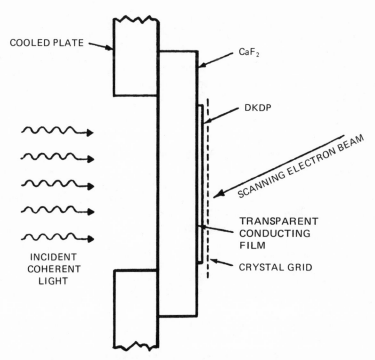

FIGURE 9.7. The electronic storage device of Figure 9.6.

Another suggested procedure using microwave holography for real-time detection of concealed weapons involves only the hologram itself.[10] We shall see later, in the section on liquid crystals (Section 9.4), that such crystals can portray microwave holograms almost instantaneously, so that if the hologram itself can divulge the existence of a concealed weapon, liquid crystals can enable such detections to be made in real time. In Figure 2.7 (the microwave hologram made by the procedure indicated in Figure 2.6), even a small section at the right of this (unreconstructed) hologram provides information as to the source of the microwaves (the waveguide at the left) through the fringe curvature. Had the waves of this figure been caused by a metallic reflecting object, a similar section of that (unreconstructed) hologram would indicate the presence and position of the source of the reflected waves.

An alternative procedure which has been suggested [10] would utilize the forward-scatter signal from the concealed metallic object, whereby a vertical, CW line antenna would radiate toward a vertical-line, liquid crystal receiver spaced a few feet away and furnished with the required hologram-generating reference wave. As the concealed weapon passed between the vertical radiator-receiver units, its forward scatter signal would become visible in the liquid crystal detector unit. It is well known that the forward scattered signal of an object is much stronger than the reflected signal from the same object.[11] Figure 1.8 shows a forward-scatter, unreconstructed, acoustic hologram; the wavelength of the sound waves used in making this record is comparable to that of 10-GHz microwaves.[1]

9.3. Microwave-Hologram Antennas

Gabor has suggested an interesting form of microwave antenna based on a hologram zone plate. The zone plate involved in this concept is not a two-dimensional one, but a three-dimensional, reflection hologram zone plate, as illustrated in Figure 3.25.

Because a reflection zone plate extends in three dimensions instead of two, it is more efficient (for the waves it reflects) than a simple, two-dimensional zone plate; the straight through, zero-order component is almost completely eliminated and only the focused (real) image is formed. Gabor suggested, therefore, that a reflection zone plate would be a useful reflector antenna for radio waves.[12] He noted that this zone plate could be made to conform to part of a sphere centered at its focal point. This would substantially reduce its coma aberration, and its beam could therefore be scanned, with a moving feed, over very large angles.

The reduction in comatic aberration for a lens or reflector having a spherical contour can be seen in Figure 9.8.[13] In Figure 9.8(a), a linear, one-dimensional section of a flat lens, similar to the lens of Figure 3.7 is shown. Here F is the true focus of the lens, so that FA and FB are equal "phase" lengths, differing in actual lengths only by an integral number of wavelengths. When the beam is to be scanned, by moving the feed to the point F^1, thereby shifting the beam by the small angle θ, the length FB is changed to F^1B. For θ very small, FA and F^1A will be approximately equal. The phase discrepancy between points A and B for the new beam position is thus given by the difference in length between FB and F^1B. This phase discrepancy in radius is

$$\Delta\phi = (2\pi/\lambda)(F^1B - FB) \tag{9.1}$$

But

$$F^1B = [f^2 + (x - f\theta)^2]^{1/2} = (f^2 + x^2 - 2fx\theta + f^2\theta^2)^{1/2}$$
$$= (f^2 + x^2 - 2f\theta x)^{1/2} \tag{9.2}$$

(neglecting higher-order terms in θ).

Expansion of (9.2) yields

$$
\begin{aligned}
F^1B &= (f^2 + x^2)^{1/2}[1 - 2f\theta x/(f^2 + x^2)]^{1/2} \\
&= (f^2 + x^2)^{1/2}[1 - f\theta x/(f^2 + x^2) + \cdots] \\
&= (f^2 + x^2)^{1/2} - f\theta x/(f^2 + x^2)^{1/2} + \cdots \\
&= (f^2 + x^2)^{1/2} - \theta x/(1 + x^2/f^2)^{1/2} + \cdots \\
&= (f^2 + x^2)^{1/2} - \theta x(1 - x^2/2f^2) + \cdots
\end{aligned} \tag{9.3}
$$

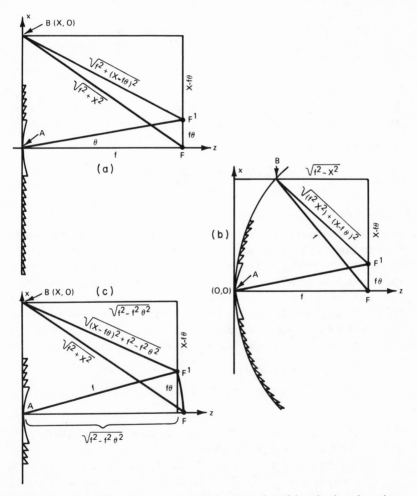

FIGURE 9.8. Aberrations in flat lenses (a) can be reduced by altering them into circular shape (b). If the focal area is also curved (c), a further reduction in aberrations results.

Therefore

$$\Delta\phi = (2\pi/\lambda)\{(f^2 + x^2)^{1/2} - \theta x[1 - (x^2/2f^2)] - (f^2 + x^2)^{1/2}\}$$

or

$$\Delta\phi = (2\pi/\lambda)[\theta x - (\theta x^3/2f^2)] \qquad (9.4)$$

The first term in the square brackets represents the expected phase shift at B due to the beam tilt, and the second term represents the undesired phase discrepancy, or coma aberration. The one-dimensional case can also be used to demonstrate how coma can be eliminated by altering the shape of the lens from flat to circular [Figure 9.8(b)]. Here the mean lens profile is curved so as to be located in a circle centered at the focus, and the phase lengths to be compared are again FB and F^1B, so that

$$
\begin{aligned}
\Delta\phi &= (2\pi/\lambda)(F^1B - FB) \\
&= (2\pi/\lambda)[(f^2 - x^2 + x^2 - 2f\theta x + f^2\theta^2)^{1/2} - f] \\
&= (2\pi/\lambda)[(f^2 - 2f\theta x)^{1/2} - f]
\end{aligned}
$$

neglecting higher-order terms in θ, and

$$
\begin{aligned}
\Delta\phi &= (2\pi/\lambda)\{f[1 - (2\theta x/f)]^{1/2} - f\} \\
&= (2\pi/\lambda)\{f[1 - (\theta x/f) + \cdots] - f\} \\
&= (2\pi\theta x/\lambda)
\end{aligned}
\qquad (9.5)
$$

which is just the correct amount of phase difference to cause the beam wave front to be tilted by an angle θ.

Inasmuch as the curved profile has eliminated coma, the remaining aberrations now consist of higher-order terms (astigmatism, curvature of the field, and distortion). A simple consideration shows that further improvement can be obtained by moving the feed approximately in a circle instead of a straight line. Figure 9.8(c) is similar to (a) except that F^1 is on a circular arc with A as its center. It is seen that the length F^1B is then

$$
\begin{aligned}
F^1B &= [(x - f\theta)^2 + f^2 - f^2\theta^2] \\
&= (x^2 - 2f\theta x + f^2\theta^2 + f^2 - f^2\theta^2)^{1/2}
\end{aligned}
\qquad (9.6)
$$

The higher-order term $f^2\theta^2$ which had to be discarded in the analysis of Figure 9.8(a) is now automatically cancelled and its contribution to higher-order aberrations is thereby removed.

A focusing reflector for narrow-bandwidth microwaves or sound waves could be constructed arraying, in three-dimensions, resonant, reflecting elements. As has been shown, such reflectors for air-borne sound waves could consist of tubes[14] a quarter wavelength long and closed at one end (Figure 9.9) or of linear rods having a U-shaped cross section[15] a quarter wavelength

FIGURE 9.9. Pipes with one end closed reflect sound waves whose wavelength corresponds to four times the pipe length.

deep (Figure 9.10). The narrow-band reflective effect of such rods is illustrated in Figure 9.11.

Two reports have appeared describing experiments with actual antennas built according to holographic principles.[16,17] These antennas are related to the "double" microwave holograms referred to in the discussion of microwave interferometric holograms (Chapter 8).

9.4. Liquid Crystal Holograms

Recently there have become available so-called "liquid crystals" which exhibit colors at precise temperatures, the color indicating the temperature. When used as thin films, they can display 20 line pairs/mm with a response time of less than a second.

C. F. Augustine of the Bendix Research Laboratories has described how these crystals can show interference patterns generated by two microwave sources.[18] The microwave field patterns are made visible almost instantaneously, with the crystals changing their color in accordance with the field strength. Figure 9.12 shows several records of such interference patterns. Figure 9.13 shows a similar photographically recorded microwave interference pattern formed by two coherent microwave sources. One of these was a point source and one approximated a plane-wave source, and the two wave sets were caused to interfere at the plane of the liquid crystal device.[19] The resulting pattern is identical to the pattern of an offset microwave zone plate.

Reconstruction of an image (either real or virtual) of the original microwave point source can be accomplished optically by photographically reducing the record and illuminating it with coherent (laser) light.

Had there been a large number of microwave point sources in the original microwave "scene," each would have formed, in conjunction with the reference wave, its own two-dimensional zone plate, and the liquid crystal pattern would have been, as in an optical hologram, a superposition of a large number of zone plates. By photographically recording, reducing, and re-illuminating this pattern with laser light, a three-dimensional, visual portrayal of the many microwave sources would have been formed. Figure 9.14 is a liquid crystal microwave hologram of the letter *L*.[20] It was made as indicated in Figure 9.15. Reconstruction of this hologram is shown in Figure 9.16.

FIGURE 9.10. *U*-shaped rods also act as reflectors for sound waves when their depth is a quarter wavelength.

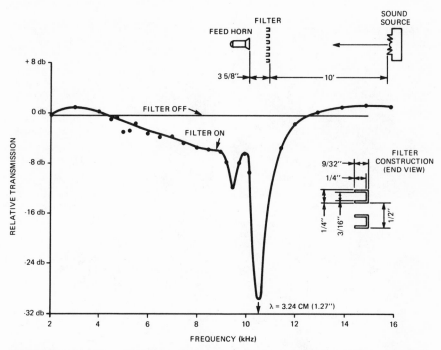

FIGURE 9.11. The reflectivity of the channel array of Figure 9.10 is strong only at the quarter-wavelength frequency.

Although very convenient for pictorially representing strong microwave fields, the liquid crystal technique requires, at the present stage of development, approximately 1 mW in.^{2} of microwave power to create a well-defined pattern. This field strength is not easy to generate for those microwave holograms which are usually of interest, namely those involving the reflection of microwave energy from rather diffusely reflecting scenes. However, with high-power microwave illumination, or as the sensitivity of the liquid crystals increases, the liquid crystal technique may become very significant in microwave holography.

9.5. Holographic Signal Processing

An interesting application of holography to the processing of radar signals was recently described.[21] The radar is of the pulse-Doppler type, transmitting a coherent train of pulses. It is able to measure, with excellent resolution, both the range and range rate of each of many targets. In the

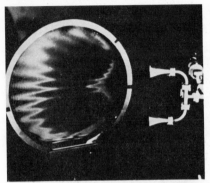

INTERFERENCE RINGS BETWEEN TWO HORNS

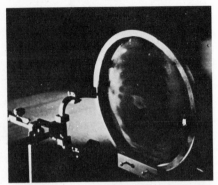

TWO HORNS ILLUMINATING REFLECTING SPHERE

FOCUS OF FRESNEL ZONE PLATE

DIFFRACTION FROM REFLECTING CYLINDER

FIGURE 9.12. Liquid crystal microwave patterns.

radar processor, a sequence of radar returns is stored as a sequence of holograms. Reconstruction of these holograms then yields the desired range and range-rate information.

In any radar, good resolution in range, which is measured by the round-trip time taken by the pulse reflection from a target, requires short-duration pulses. On the other hand, good range-rate resolution, determined by measuring the Doppler frequency shift of a signal returning from a target, requires a long-duration pulse. To satisfy these two opposing requirements, a long train of many very short pulses is employed as the transmitted radar "pulse" signal. The signal remains coherent over the length of the pulse train. As shown in Figure 9.17 the burst consists of a train of N pulses equally spaced T seconds apart, with each pulse having a duration of t sec. The pulse duration t determines the range resolution, and the burst duration NT determines the Doppler frequency-shift resolution (equal to $1/NT$ Hz). In the middle part of Figure 9.17 it is seen that each radar transmission elicits a return from each of the four targets. The delay between a transmitted pulse and the time of the corresponding return from one of the targets establishes the target's range with the required resolution. The range *rate* of each target is obtained by feeding the entire sequence of N returns from the target into a spectrum analyzer and noting the Doppler frequency shift (the bottom line of Figure 9.17). Since such a spectral analysis must be performed for each

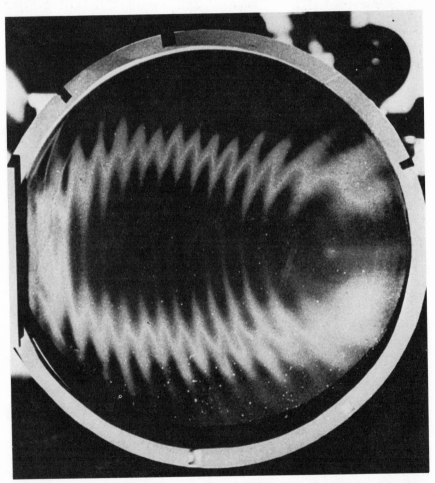

FIGURE 9.13. A liquid crystal pattern formed between a microwave point source and a plane-wave source. This pattern, a microwave hologram, is a zone-plate, which when greatly reduced in size and illuminated with laser light can reconstruct virtual and real images of the original microwave point source. Because it is offset, the straight-through (zero-order) component and the two images can be separated.

FIGURE 9.14. Microwave hologram of the letter *L*.

element of range of interest, many spectrum analyzers are normally needed. In the new technique, simple holographic procedures accomplish the same task, and thus eliminate the need for the many analyzers. The holograms, acting as optical gratings, in effect are the spectrum analyzers.

In the first step in holographic processing, the radar returns are stored holographically, as shown in Figure 9.18. A collimated beam of laser light is split into reference and signal components, and these are later recombined to form an interference pattern on a surface of photographic film. In the upper path the signal beam is generated by passing laser light through an ultrasonic light modulator in which the radar return is traveling (at the velocity of sound). The lower path directs a second, tilted, plane-wave, laser reference beam onto the film surface. The signal in the aperture of the ultrasonic light modulator is the reflected signal obtained from the several targets that are interrogated by one transmitted radar pulse of the train. When the entire radar return has moved into the ultrasonic light modulator, the laser light is briefly turned on to provide the required hologram exposure time. This one hologram occupies one narrow vertical strip on the film. After each individual pulse of the pulse train has thus been photographically recorded, the film is moved sideways, i.e., perpendicular to the plane of the figure, so that a sequence of *N* radar returns is recorded as *N* holograms side by side on the photographic film.

This series of holograms forms, for each target, an optical grating whose tilt is a function of the range rate of that target. After the film sheet of many side-by-side holograms is developed, it is again illuminated by the original reference wave. The holographically reconstructed configuration in shown in Figure 9.19. It shows two information-bearing areas outlined in the output or frequency plane. Either of the two areas contains all the range-Doppler information.

An expanded view of one of these areas is shown in Figure 9.20. Each of the four bright spots of light corresponds to one of the targets that was interrogated by the N transmitted radar pulses. The horizontal position of the spot is indicative of the target's range, and the vertical position is indicative of the target's range rate.

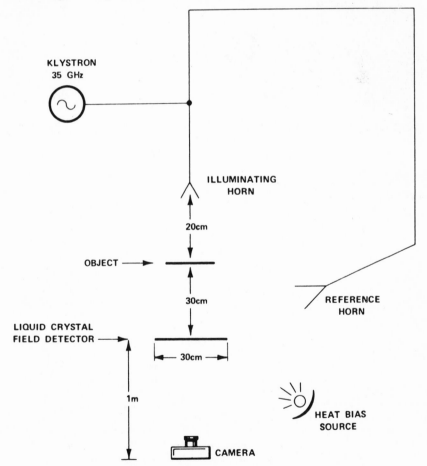

FIGURE 9.15. Procedure for making the hologram of Figure 9.14.

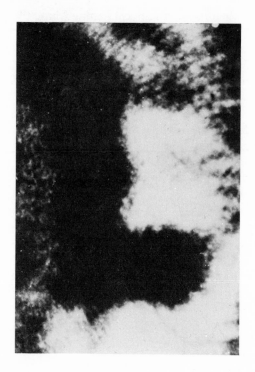

FIGURE 9.16. Reconstruction of the hologram of Figure 9.14.

9.6. The Synthetic-Aperture Concept

One of the most extensive uses of nonoptical holograms has occurred in the microwave radar field in the form of synthetic-aperture radar. Because of the outstanding success of such hologram radars, similar hologram concepts are beginning to be examined in the sonar field. In 1971, shortly before the announcement of his award of the 1971 Nobel Prize in Physics, for holography, Gabor commented:[22]

> Unknown to me, a most interesting branch of holography was developing from 1965 onwards at the Willow Run Laboratory attached to the University of Michigan. It was holography with electromagnetic waves, and reconstruction by light, which was called "Side Looking Radar" or "Synthetic Aerials." It was classified work; the first publication by Cutrona, Leith, Palermo and Porcello occurred in 1960. Reconstructions of the object plane by illumination with a monochromatic mercury lamp were of impressive perfection. So, curiously, in the first 12 years, the aim of holography was the reconstruction with light of electron or X-ray records, with wavelengths about 100,000 times shorter than light, and reconstruction with light of electromagnetic holograms with wavelengths 100,000 times longer.

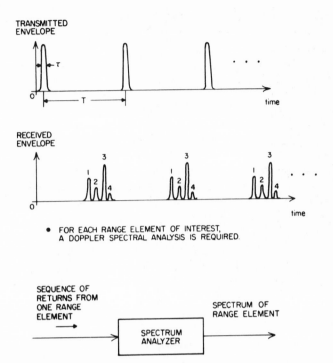

FIGURE 9.17. Outgoing radar signal comprising a train of short pulses (top). Below it is shown the received signal with four targets present. The spectrum analyzer at the bottom determines the Doppler shift for a given target range.

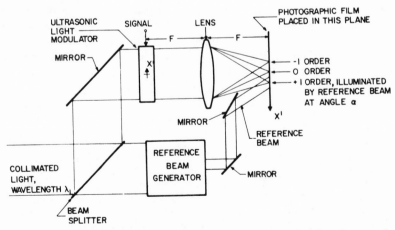

FIGURE 9.18. Procedure for generating hologram for radar signal processing.

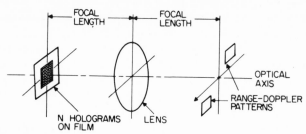

FIGURE 9.19. Reconstructing the hologram made as shown in Figure 9.18.

Because Emmett Leith was one of the principal contributors to synthetic-aperture development, and because he also was, with Upatnieks, the first to use the laser in holography, Gabor has commented:[23] "Emmett Leith arrived at holography by a path just as adventurous as mine was. I came to it through the electron microscope, he through side-looking, coherent radar."

Thus, it is a matter of record that whereas Gabor's brilliant conception of optical holography in 1947 lay almost dormant until Leith and Upatnieks initiated in 1963 what has now become a massive world-wide development effort on it, the electromagnetic form of holography received extensive attention all through the 1950's. As noted by Gabor (above) the air-borne form of synthetic-aperture radar was pioneered by L. J. Cutrona and his group at the University of Michigan,[24,25] and sizeable support was given it in classified projects in numerous laboratories across the U.S. Indeed, it is likely that this early classified program was in large part responsible for the extreme rapidity of growth in the development of true optical holography from 1963 on, particularly at the University of Michigan.

As University of Michigan Professor William G. Dow has noted, the involvement of numerous University of Michigan professors in this classified work later enabled them to teach their students the subject of optical processing and holography in a much more timely and effective way.[26] Actually, the relation of coherent radar to holography was not appreciated until rather recently,[27,28,29] but Leith, a key contributor to this radar development, instinctively applied the side-looking radar technique of offset beams,[24] i.e., the two-beam recording technique, in his first experiments in the optical

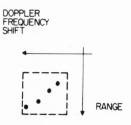

FIGURE 9.20. An enlarged view of a reconstructed hologram.

holography. The success of synthetic-aperture radar led this author to consult with Cutrona in 1957 on the possibility of applying this radar concept to sonar devices. Unfortunately, when the original synthetic-aperture technique is applied to sonar, the low propagation velocity of sound somewhat limits its usefulness; however, as we shall see, new procedures have recently been described which can materially extend the use of synthetic-aperture procedures in sonar applications.[30]

For simplicity we define hologram radar or sonar systems as those which employ sound waves or radio waves having extremely high temporal coherence, and which utilize that coherence with an equally coherent reference wave to generate (e.g., by synchronous detection) a recorded interference pattern (a hologram) for later processing (reconstruction) of the reflections from objects. As noted, many of the early synthetic-aperture developments took place in the radar field during the early 1950's, with acoustic versions following some years later. In that early work and until quite recently, the synthetic-aperture concept was considered as a Doppler detection process, and even in the classic 1966 paper of Cutrona, Leith, Porcello, and Vivian,[25] no mention of holograms was made. In 1967, the relation of this technqiue to holography was analyzed at length,[27] and since then numerous publications have appeared discussing this connection,[28,29,31–36] with one (Reference 29) stating that "the holographic viewpoint appears more flexible than the communications theory or cross-correlation viewpoint and has led to designs which are not easily explicable from the latter viewpoint."

The earliest *published* paper on hologram radar (as defined above) was by G. L. Rogers.[37] In those 1955 experiments, the reflector was moving and the radar was stationary. Hologram concepts were employed, including the use of a "signal taken from the transmitter's master oscillator, suitably attenuated, and fed into the receiver, where it combines with the downcoming (reflected) signal in accordance with the well-known 'coherent demodulator' technique."[37] Rogers, in these experiments, generated a one-dimensional image of a Dakota aircraft, which had purposely been sent up, so as to permit him to recover, "by diffraction from one of the aerial records" such an image. In 1966, G. L. Tyler[38] discussed a bistatic form of hologram radar, noting that his bistatic-radar mapping scheme corresponded to the process of making a (radar) hologram and "playing it back." The widest use, presently, of hologram (synthetic-aperture) systems involves air-borne radar, although synthetic-aperture acoustic (ultrasonic) systems are now being examined.[39]

In a synthetic-aperture radar system,[25] an aircraft moving along a very straight path continually emits successive microwave pulses. The frequency of the microwave signal is very constant (the signal remains coherent with itself for very long periods). During these periods the aircraft may have traveled several thousand feet, but because the signals are coherent, all the many echoes which return during this period can be processed as though a single antenna as long as the flight path had been used. The effective antenna length is thus quite large and this large "synthetic" aperture provides records having extremely fine detail.

The photographic record of the echoes received by such a coherent radar is a form of hologram, with the microwave generator which provides the illuminating signal also providing a reference wave. The reflected signals received along the flight path are made to interfere with this reference signal (by synchronous detection), and the complex interference pattern thereby generated and photographed is a form of hologram.

The method of operation is shown in Figure 9.21. For simplicity, only one reflecting point is shown. Waves returning from this point have spherical wave fronts, whereas the oscillator reference signal acts like a set of plane waves perpendicular to the path of the airplane. The received signal is combined with the coherent reference signal and amplified to intensity-modulate a cathode-ray-tube trace as shown in Figure 9.22. Each vertical line in that figure thus plots signals received from all range points, with the points at greater range being recorded near the top of the vertical trace. As the airplane moves along and new pulses are emitted, the film is indexed to record a new set of returns on a new vertical line.

For the case of only one reflecting point at fixed range, the upward-moving cathode-ray beam would, for every pulse, be modulated only at that one point in range, and the result would be a single horizontal line of recorded echoes. But this line is not continuous. The returning waves are circular, and

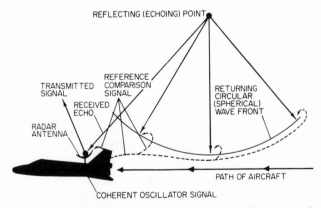

FIGURE 9.21. In a synthetic-aperture radar, echoes from the reflecting point are received over a considerable interval and assembled by a holographic method that creates interference patterns between the return signal and a reference sampled from the transmitted signal. The result is a synthetic-aperture high-resolution antenna whose length is that of the flight path.

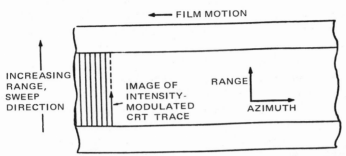

FIGURE 9.22. A hologram record in a synthetic-aperture radar is formed by photographing repeated intensity-modulated cathode-ray-tube traces.

as the slant azimuth range from the aircraft to the reflecting point changes, the combination of return and reference waves produces successive constructive and destructive interference. At the greater slant angles, this succession of in-phase and out-of-phase conditions occurs rapidly, whereas when the aircraft is practically abreast of the point, it occurs slowly. The resulting record is a one-dimensional zone-plate hologram, as shown in Figure 9.23. When illuminated by laser light, as shown in Figure 9.24, it reconstructs the reflecting point just as a hologram does.

The range of each of a multitude of reflecting objects is recorded and each object is reconstructed by illuminating the hologram with a laser, as shown in Figure 9.25. Indicated in the figure are two reflecting points which are displaced appreciably in range and slightly in azimuth. All reconstructed images fall on a tilted plane as shown. The tilt of this plane is determined

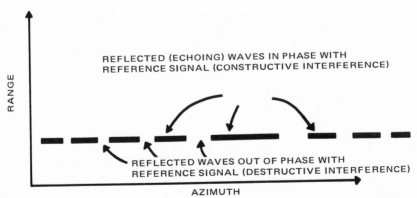

FIGURE 9.23. When a photographically stored hologram of a single point is made, the recorded signals form a one-dimensional zone plate.

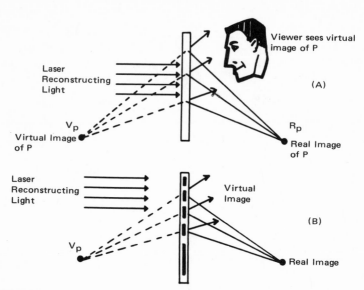

FIGURE 9.24. This sketch shows the similar diffraction effects in (A) a true holo-gram of point source P; (B) a one-dimensional microwave hologram (zone plate) of single radar reflecting point P.

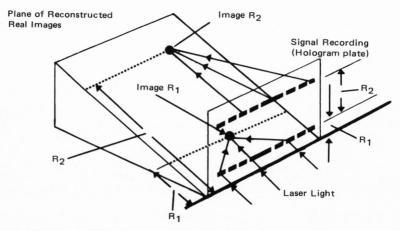

FIGURE 9.25. Reconstruction by laser light of a side-looking radar hologram formed by two reflecting points.

FIGURE 9.26. A synthetic-aperture (hologram) record.

by the amount of radar vertical tilt; for a radar having an antenna which illuminates the ground at almost grazing incidence this tilt is very small.

A typical microwave hologram as generated by a side-looking radar is shown in Figure 9.26 and an enlarged portion of this radar hologram is shown in Figure 9.27. In the latter figure, a prominent one-dimensional zone plate is seen near the bottom of the central blank area. As in optical holograms, this assemblage of microwave holograms does not identify the object or area in view. Nevertheless, the *processed* hologram yields photographs of

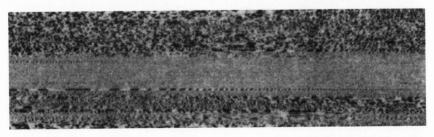

FIGURE 9.27. A portion of the hologram record of Figure 9.26. The one-dimensional zone-plate pattern of a particularly distinct reflecting object is seen at the lower portion of the blank area.

extremely good detail. One of the early examples, the city of Washington, D.C. and its environs, is shown in Figure 9.28. During 1971 and 1972, the entire Amazon River Basin was similarly recorded; Figure 9.29 is an example of a (reconstructed) record made during this mapping project. Figures 9.30 and 9.31 are two further examples of the detail possible from synthetic-aperture radars.

When coherent radar was first described, many could not understand how its high resolution could be maintained for all objects. Because of the great length of the synthetic-aperture antenna, reflecting objects recorded by it are in its *near* field, not in the distant, or Fraunhofer, region, where most

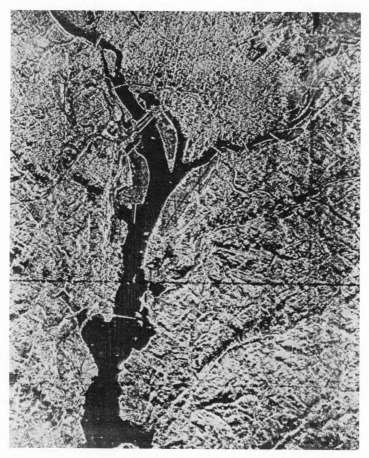

FIGURE 9.28. Proper optical processing of synthetic-radar holograms provide high-resolution detail of the terrain flown over by the aircraft. This is a record of Washington, D.C. (courtesy of L. J. Porcello).

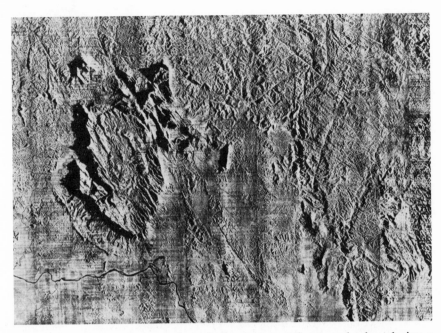

FIGURE 9.29. A reconstructed synthetic-aperture-radar record taken during a 1971–1972 survey of the Amazon River Basin (courtesy of Homer Jensen, Aero-Service).

radars and sonars operate. For a near-field reflecting object, radars or sonars having large-aperture receiving arrays can only achieve maximum efficiency and resolution when the time delays inserted in the receiving-element circuits are adjusted so as to cause the element positions to correspond to being arrayed along the arc of a circle centered on that object. Yet, for other near-field objects, at other ranges, all of these delays would have to be changed so as to correspond to arcs of other circles, centered on these other points. How the mysterious optical processor could cause all points in the near field, at any distance and at any angle, to be in sharp focus, was beyond comprehension to many. When it is recognized, however, that the radar record is a hologram, the answer becomes quite clear. In holography, each small light-reflecting point generates its own plate, and each zone plate causes coherent laser light to be reconstructed exactly at the point in space from whence the light emanated. Similarly, a synthetic-aperture radar captures photographically the curved wave fronts emanating from reflecting points by combining them with a reference wave and thereby generating (one-dimensional) zone plates. Later, as in an optical hologram, the coherent light used in the reconstruction process acquires, through diffraction by the zone plates, the properly curved wave fronts to concentrate the light at points corresponding to the reflecting points in the original landscape. Just as an optical hologram causes each point of a three-dimensional scene to be brought into sharp focus no matter what its distance from the photographic plate, so the synthetic-aperture-hologram record is responsible for the good focus of all of its reconstructed points. This is shown quite well in Figure 9.29. As noted above, this figure was made during a recent synthetic-aperture-radar survey, of the Amazon River Basin (3 million square kilometers), conducted by Aero-Service Corporation of Litton Industries under the supervision of Homer Jensen.[40] It is a print of the area near the town of Esmeralda in the State of Amazonas in Venezuela. The river in the lower-left-hand corner is the upper Orinoco. The radar employed was an X-band (10GHz) radar made by the Goodyear Aerospace Corporation, and the Venezuelan title for such a record is "Mosaico de Radar." The object of this survey was to provide accurate maps of the area, not just a reconnaissance or surveillance of it. For this special requirement, Aero-Service made several additions to the processing equipment, including an anamorphic printer to correct for scale variations, and a special optical mask to correct for tonal effects (I am indebted to Homer Jensen for these details).

Even in fairly recent discussions of synthetic-aperture radar, including

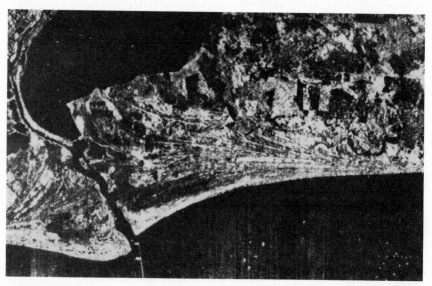

FIGURE 9.30. Contrasting aerial view of Port Arthur, Tex. shows water as black areas and ships and oil rigs as white specks (courtesy of Goodyear Aerospace Corporation).

FIGURE 9.31. A hologram reconstruction of the city of Phoenix, Ariz. (at the middle right). The block patterns at the top and at the left are irrigated farms. The geological features are clearly seen at the top right, at the bottom right, and at the bottom left. The white area at the left is blocked out for security reasons (courtesy of Goodyear Aerospace Corporation).

several which note the relation to holography, the very early Doppler concepts are still stressed.[32] In true holography, there is no Doppler, so that once the holographic viewpoint is fully embraced, there is no need to consider Doppler effects.[31] So, as we shall see, the way is then open for consideration of other forms of coherent systems, such as stationary (Doppler-free) radars (and sonars) and others.[28,31,33-35]

Because the usual optical imaging process employs lenses or paraboloidal reflectors, only one plane section of the image field can be recorded in truly sharp focus; all other planes are *out* of focus in varying degrees. Ordinary sonars and radars obey similar optical laws. However, this focusing problem is absent in both holography and synthetic-aperture systems. As we observed above, each point in the hologram scene (or each reflecting point in the radar or sonar field) forms its own zone plate, and each such recorded zone plate then causes coherent light to be reconstructed (i.e., focused) at the proper (equivalent) point in space.

In synthetic-aperture systems, the length of the recorded zone plates depends upon the range of the reflecting objects which generates them. This

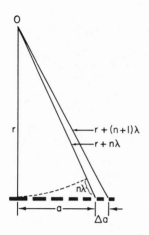

FIGURE 9.32. The geometry of a half zone plate.

effect is illustrated in Figure 9.32. In this figure, Δa is the minimum resolvable fringe spacing, and this is determined by (i.e., it is closely equal to) the horizontal aperture of the moving radiator (receiver). For the radar case this radiator is the aircraft's side-looking antenna. It is seen from Figure 9.32 since $a^2 \approx 2rn\lambda$, and $a = r\lambda/\Delta a$, that when $\lambda = 0.1$ ft and $\Delta a = 20\lambda$, the half zone-plate length (the half synthetic-aperture, a) is 5000 feet at a 100,000-ft range but only 50 ft at a 1000-ft range.

This variation of synthetic-aperture size with range results in the resolution of such systems to be independent of range. This is illustrated in Figure 9.33, showing the focusing action of a zone plate. The heavy lines indicate the envelope (to the first null) of the wave energy which is focused (diffracted) by a zone plate Z having an aperture $2a$. Assuming that the energy concentration is diffraction-limited, the azimuthal resolution, R_a, is approximately equal to $\lambda r/a$, and the range resolution, R_r, is approximately equal to $\lambda r^2/a^2$. These results show that when $r = 10.000$ ft, $\lambda = 0.1$ ft, and $\Delta a = 20\lambda$, R_a is only 2 ft (i.e., it is equal to the antenna size). Thus, for these parameters, the

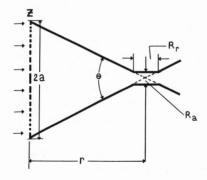

FIGURE 9.33. The focusing action of a zone plate.

azimuthal resolution is 20 times better than the range resolution. Also, when the range r is increased (or decreased), the zone plate, Z, of Figure 9.33 becomes larger and moves to the left (or becomes smaller and moves to the right), with the angle θ and also the values of R_a and R_r remaining unchanged. We note that for the example picked earlier, when the half synthetic aperture, a, was 5000 ft, the ratio of the length of the zone-plate aperture (10,000 ft) to the azimuth resolution (2 ft) is 5000. For the situation where Δa extends over only one wavelength, the range and azimuth resolutions become equal. Under these conditions good range resolution would be acquirable without the use of pulses, i.e., it would be provided by the (superimposed) zone plates alone.

9.7. Stationary Synthetic-Aperture Systems

Until recently, the usual form of synthetic-aperture (hologram) system has involved a moving transmitter and receiver, but, as in true holography, much broader uses of coherent sonar and radar concepts can be made, particularly when large apertures are involved. Such sonar or radar arrays could take the form of extremely long linear arrays of independent receivers. Other forms include crossed arrays (Mills crosses) and square arrays. In holography even a small portion of the hologram is able to reconstruct the full image. Similarly, in stationary coherent sonars or radars, retaining only the end sections of long linear arrays, or the four corner sections of square arrays (thereby maintaining maximum resolution) should be satisfactory in many situations.

Perhaps the simplest form of a stationary coherent radar or sonar would consist of a very long linear array of elements, each element being identical to the antenna carried by the aircraft in the usual side-looking radar. In operation all elements would first transmit, simultaneously and in synchronism, a short pulse of coherent microwaves, following which all units would, as in an ordinary radar or sonar, act as receiving elements. A certain amount of the coherent oscillator signal would continually be fed to each of the elements (for example, through the procedure of coherent detection), thereby providing the holographic reference signal. The returning, reflected echo signals would interfere at each receiver unit with this reference wave, so that along the entire array a wave interference pattern would exist. This pattern could be photographically recorded, for example, through the use of a cathode-ray tube, having, instead of a single beam, a large number of beams (the number corresponding to the number of elements in the receiver array), with the beams intersecting the luminous face of the tube in a single horizontal line, and all moving upward

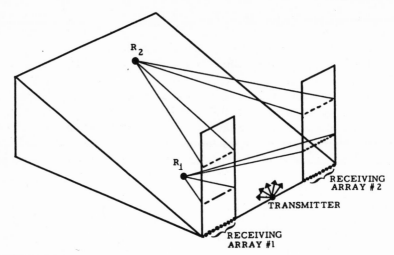

FIGURE 9.34. A stationary, line-array, hologram sonar can also be effective when only a portion (e.g., the two ends) is employed.

together. The intensity of all beams would, as in Figure 9.22, be modulated by the output signal from each elemental array antenna. As in Figure 9.22 (and in the radar records of Figure 9.26), range would be plotted vertically, but instead of the vertical lines being recorded sequentially by one beam, all would be recorded simultaneously by the many, upward-moving beams. A photograph made of the picture generated on the face of the tube would thus be completely equivalent to the radar record of Figure 9.26. When processed in the same manner as the radar record, a reflecting object would similarly generate a one-dimensional zone-plate interference pattern (as in Figure 9.23), and the reconstruction of the entire, complicated, record would similarly provide a picture of the area located in the field of the radar or sonar. We shall later discuss an alternative use of such a multibeam cathode-ray tube in synthetic-aperture radar or sonar, which has recently been described, and which extends the system's maximum-range capability.

It was seen in Figure 3.18 that focusing devices such as paraboloidal surfaces or zone plates can achieve a focusing effect even if only a portion of their focusing surface is utilized. Because a hologram is a form of zone plate, it, too, can achieve certain of its effects if only a small portion of the original hologram is used in the reconstruction process. We noted this in connection with Figure 5.15, which portrays this property for a hologram zone plate; both the full area $ABCD$ and its smaller portion $A'B'C'D$ generate real and virtual images. Similarly, the linear array just discussed need not comprise the entire length; only the end sections need be used, as shown in Figure 9.34. In this figure the transmitting unit is separate.

9.8. Forward-Scatter Hologram Radars

Synthetic-aperture concepts can be extended to a forward-scatter radar or sonar, i.e., to a coherent system modeled after the original Gabor hologram arrangement.[41] In this arrangement, the area or targets of interest would be located between the transmitter and a receiving array, i.e., the transmitter would be placed at 180° from the receiver (relative to the illuminated area of interest). As we shall see, such an arrangement would take advantage of the rather high-forward-scatter signal diffracted by objects located between the transmitter and receiver.[11] The receiver could either be one long, linear array of many elemental receivers, or two end sections of that array. In the hypothetical case where only one point-scatter is located between the coherent source and the receiving array, the interference pattern between the scattered signal and the coherent background signal would again be a zone plate, and the receiving array would intercept a linear section of it (a one-dimensional zone plate). The arrangement is seen to be similar to the method of making an optical zone plate shown in Figure 9.35, with the (distant) transmitting source generating approximately plane waves at the receiver array, and the target, corresponding to the pinhole of that figure, causing spherical waves to

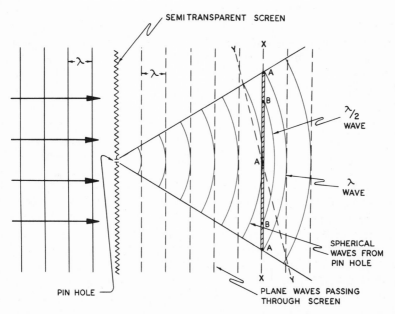

FIGURE 9.35. Plane and spherical waves combine to make a zone plate photographically.

be scattered toward the array. By photographically recording the individual outputs of all of the receiving-array elements as one photographic line (comparable to one line of a synthetic-aperture-radar record), the position both in range and azimuth of this scatterer could be determined by processing the one-dimensional photographic zone plate with laser light.

In the usual case in which many scatterers are present, a multiplicity of zone plates will be generated, and on the photographic record these would all be superimposed. As in a hologram, however, positional information on all of the scatterers could still be retrieved through illumination of the photo-graphically recorded single-line interference pattern with laser light. To make the photographic line record, the receiver-element outputs could be sampled sequentially, and the resulting signal could intensity-modulate a synchronous-ly moving cathode-ray-tube beam with the light signal then being recorded photographically. It has been shown that in the forward-scatter case, the scattered signal is quite strong.[11,14]

In most sonars the target is illuminated by a transmitter, and the echo returns to a receiver placed at the transmitter location. For such arrangements, the signal scattered back from an object is almost always extremely small. On the other hand, the signal scattered or diffracted in the forward direction is usually much larger. This strong signal in the forward direction can be explained by Babinet's principle (Figure 9.36): An aperture or hole in a screen upon which a plane wave is falling creates the same diffracted field in the forward direction as that created by a disk having a size equal to the screen aperture. Now the forward lobe in the case of an aperture is identical with the beam of a radiator having the size of the aperture and radiating a plane wave. It is evident that this forward lobe is always stronger than the echo reflected back from an object except in the case where the object is perfectly flat.

Babinet's principle applies strictly only for an infinitely thin screen and for an infinitely thin object replacing the aperture. Thick, irregularly shaped objects are difficult to treat analytically, and because of this one cannot conclude theoretically that the forward lobe exists in equal magnitude for a thin, flat disk or for a thick, irregularly shaped object having the same shadow area. Straightforward reasoning suggests, however, that the lobes should be similar, since, for objects large with respect to the wavelength, the energy striking the front face of the object is not too instrumental in creating the forward lobe.

One technique for demonstrating the forward-scatter lobe is an acoustic one. Figure 9.37 shows the measurement techniques employed in these tests. In a free-space room, loudspeaker *A*, radiating a steady, single-frequency tone, was placed so that sound could reach microphones *B* or *C*. With no object present, a signal of proper amplitude and phase was added to the two microphone circuits so as to cause complete cancellation of the directly

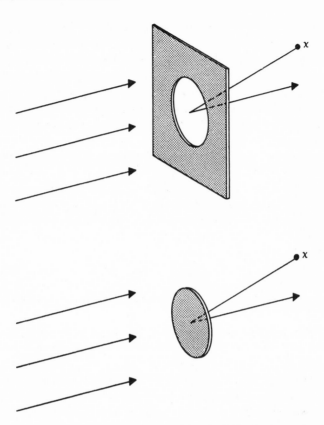

FIGURE 9.36. Babinet's principle points out the similarity between the diffracted wave pattern of a circular disk and the same size circular aperture.

received signals. When an object was placed in the position shown, the reflected signal was observed in microphone *C* and any forward-scatter signal was observed in microphone *B*.

When a rigid disk was placed in position equidistant between loudspeaker *A* and microphone *B* the signals observed in the two microphone circuits were equal. If the disk was then rotated around a vertical or horizontal axis, the echo at receiver *C* was reduced markedly, but very little change was observed in the forward lobe until the projected area of the disk was reduced an appreciable amount. When the object was a sphere, the forward-scatter signal was very closely equal to that of the disk of the same cross section, but the signal at receiver *C* was greatly reduced over that produced by the oriented disk. A sphere of sound-absorbing material also created an equal forward-scatter signal, but the back-scatter signal was, of course, extremely small. Directional

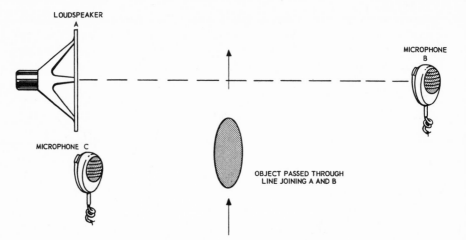

FIGURE 9.37. An experiment to measure the forward-scatter pattern of a disk.

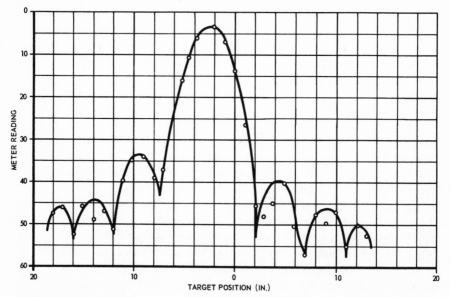

FIGURE 9.38. With the direct signal balanced out, the right-hand microphone of Figure 9.37 generated this record of the energy scattered in the forward direction. The similarity to the beam created by plane waves passing through an equivalent circular aperture is evident.

patterns of the forward lobes of the various objects were measured; they corresponded quite well with the calculated values (assuming the object to be equivalent to an equal size hole in a Babinet screen). The pattern of a disk is shown in Figure 9.38, and we saw in Figure 1.9 a visual sound portrayal pattern of the forward lobe. In the forward-scatter (bistatic) hologram sonar or radar, the transmitter could operate continuously (i.e., it would not have to be pulsed), and both range and bearing (azimuth) could be acquired by the array.

9.9. Bistatic Hologram Radars

Continuous-wave bistatic systems (radars or sonars) are often employed to provide information on moving targets. Because such targets usually generate signals at the receiver which closely approximate a zone-plate pattern, hologram procedures offer a significant improvement in the signal-to-noise ratio in such radars.[42] When an aircraft flies across the line joining a transmitting station and a receiving station, the signal reflected from the aircraft combines at the receiver with the direct transmitter signal, and this generates interference effects which cause the (combined) received signal to vary periodically, rapidly at first, then slowly, then rapidly again. This effect is often observed on home television sets. A time–frequency plot of such changing-frequency signals is shown in Figure 9.39 for a number of passing aircraft. Comparable records would be generated by targets moving between

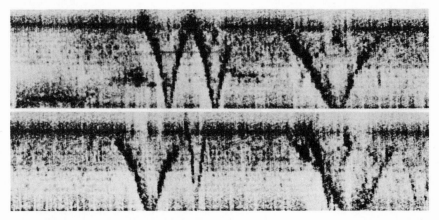

FIGURE 9.39. Frequency *vs.* time records of the interference effect caused by aircraft flying over a radio transmission link.

a CW, underwater, sound radiator and a similarly placed acoustic receiver. For both the radar and sonar case, constant-velocity targets generate signals which have the time–amplitude pattern of a zone plate. Accordingly, coherent (hologram) optical procedures would provide an improvement in the signal-to-noise ratio at the receiver. To indicate the similarity between such transmission-path interference signals and synthetic-aperture records, two situations are shown in Figure 9.40. As in Figure 9.21, the left side of the figure portrays the process of recording a single reflecting object O on a synthetic-aperture-radar record. The right-hand portion of Figure 9.40 indicates the arrangement involved in generating the signals of Figure 9.39; the CW signal from the transmitter acts (at the receiver) as the reference signal, and there is a moving reflector P rather than a moving transmitter and receiver. It is thus evident that the individual V curves of Figure 9.39 could have been recorded photographically, as one-dimensional zone plates. For the targets traveling at uniform velocity along a straight line, such hologram procedures would provide a signal-to-noise gain equivalent to the signal-to-noise improvement obtained in moving synthetic-aperture systems.

9.10. Passive and End-Fire Radars

The synthetic-aperture concept can also be applied to passive forms.[43] We consider again a long, multielement receiving array and assume that there is a target present which is radiating a strong, single-frequency, highly coherent, signal (a radio signal or an acoustic one). We assume further that one of the array elements receives this signal, amplifies it, and feeds it as a

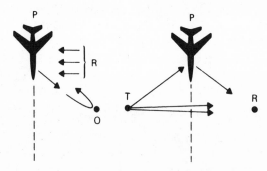

FIGURE 9.40. The recording, by a synthetic-aperture radar, of echoes from a single reflecting point (left), is similar to the recording of the signal received in a radio transmission link when it is flown over by an aircraft (right). The direct signal from T to R (right) corresponds to the reference wave R (left).

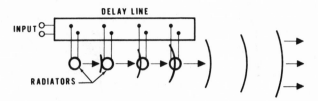

FIGURE 9.41. An end-fire array results when radiators are placed in a row and energized, not in phase, but in such a way that proper phase addition occurs in a given direction. In the illustration, this direction is to the right.

reference signal to all of the other elements of the array. These elements will, however, also be receiving the radiated signal from the target directly, so that an interference pattern will again be generated along the array. This pattern can be photographically recorded and used to reconstruct the location of the target in azimuth and range. In this arrangement there is no radiation outward from the system, so that is constitutes a *passive* system, as contrasted to a *transmitting*, or active, system.

The usual form of synthetic-aperture (hologram) system is the side-looking version (Figure 9.21), which generates a synthetic aperture corresponding to a linear, broadside array. Recently, the concept was extended to a form in which the synthetic-aperture resembles a linear *end-fire* array.[44] When a series of radiators are energized by a single source and delay is introduced between the individual radiating elements, the radiated waves can be made to add along the line of radiators, i.e., an end-fire pattern, as shown in Figure 9.41. The photographically recorded interference pattern that is produced by combining the signals echoing from a target which is ahead (for example, on the forward course of a ship) with a coherent reference wave is a *uniformly* spaced, varying-density pattern, rather than the nonuniformly spaced zone-plate pattern of the side-looking case. The pattern can be looked upon as a one-dimensional grating, and to reconstruct the object (the target) a lens is employed to convert plane laser light waves diffracted by this photographic grating into circular waves converging at a focal point. The light concentration thereby is equivalent to that obtained with a one-dimensional zone plate, and a correspondingly high array gain (with its accompanying high signal-to-noise ratio) is realized. As in the side-looking case, pulsed transmissions would normally be employed to provide range information on the reflecting objects of interest.

Whereas the usual synthetic-aperture radar (or sonar) operates from a moving platform, the same end-fire concept can be applied to a *stationary* (active) system whereby a small transmit–receive transducer would effectively be provided with a large, synthetic gain against targets moving toward it or past it at uniform speeds.[45] Figure 9.42 indicates the procedure for imparting

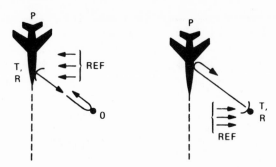

FIGURE 9.42. The synthetic antenna gain achieved in an air-borne synthetic-aperture radar (left) can be duplicated for a stationary radar against an aircraft target moving in a straight line at uniform speed (right).

synthetic gain to a stationary radar antenna. At the right is a moving reflector *P*, rather than the moving transmitter and receiver of an air-borne synthetic-aperture radar (as shown at the left). When a reference signal is introduced, the combined signal is, for identical aircraft motion and with a similar-moving film technique, the same for both cases. The concept is of course also applicable to sonars. For underwater targets traveling at uniform velocity along a straight line, the signal-to-noise gain thereby achieved at the stationary sonar would be exactly equivalent to the signal-to-noise improvement of the moving synthetic-aperture sonar.

We noted above that end-fire gain could be imparted to a moving synthetic-aperture system. This technique is also possible for a stationary-system sonar when uniform-velocity targets are moving directly toward it. Since the synthetic gain thereby achieved would be generated only by moving targets, reverberation effects would be minimized, and the long-range capability of the sonar should be markedly improved.

9.11. Synthetic-Interferometer Radars

Recently, L. C. Graham of Goodyear Aerospace Corporation described an interferometric synthetic-aperture (hologram) radar for topological mapping which provides, in addition to the extremely fine resolution in azimuth (and range) of the standard synthetic-aperture radar, information on a third, vertical, dimension.[46] In contrast to the *holographic* interferometry described in Chapter 8, the interferometry technique employed in this system uses two receiving antennas, widely spaced in the vertical direction, with the two received signals added vectorially. The receiving pattern of such a combined-spaced pair contains numerous closely spaced minima (as in Figure 1.15). These minima, for an accurately matched antenna pair, are extremely deep, and, at their large depths, extremely narrow. This narrowness provides very accurate elevation (third dimension) information, but at these large depths, the amplitude of the received signals is low, and the signal-to-noise ratio is therefore lower in the minima than at the peaks. The high synthetic gain, resulting from the large (linear) synthetic *aperture* generated by the hologram process, enhances the signal-to-noise ratio, and it can be further improved by increasing the transmitter power, or by employing pulse compression, a subject we shall discuss in Chapter 10. Examples are given in Reference 46 which show that the final processed (reconstructed) images for this three-dimensional radar provide far more realistic pictures when the terrain surveyed contains mountains or hilly areas. (Figure 9.43).

9.12. Radar Range Acquisition by Holography

Both the usual radar systems, and those utilizing the synthetic-aperture technique just described, employ pulses for measuring the range to a target, determined by the time taken by the pulse in traveling from the transmitter to the target and back to the receiver. Once it was recognized that the synthetic-aperture procedure is a form of holography, other forms of hologram radars became apparent, including, as we have just noted, end-fire (synthetic-gain) systems, bistatic systems, and various stationary extensions of moving synthetic-aperture systems. It was in the discussions of such stationary holo-gram radar systems that the concept of range acquisition by holographic means was first described.

We described earlier one possible example of a stationary radar comprising a long linear array of transmit–receive elements with the received signals modulating a plurality of upward-moving cathode-ray-tube beams. We noted that an alternative arrangement could be that of Figure 9.35, in which a separate transmitting unit is utilized. The notion that range can be acquired by holographic procedures, without the use of pulses, dates back to 1968.[31] In 1973, the first published experimental data demonstrating valid-ity of the concept was reported.[47] In these experiments, a situation was simulated on a computer in which an S-band (10-cm)-wavelength, CW (*not pulsed*) radar having an antenna aperture of 10 m was found to exhibit a response at the proper range of 28 m and practically zero response at other ranges (0–200 m). This particular hologram radar is being constructed for the purpose of measuring the thickness of ice layers. A further explanation of the hologram-matrix concept of the 1973 work appeared about a year later.[48]

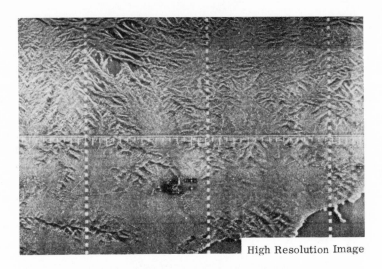

High Resolution Image

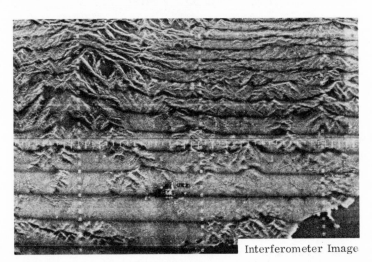

Interferometer Image

FIGURE 9.43. An interferometric synthetic-aperture image (bottom) provides an improved altitude portrayal of the terrain (courtesy of Goodyear Aerospace Corporation).

In that system, the phases of the array elements were adjusted so as to generate a circular wave front. In the procedure described earlier in this chapter, no pulses would be used, so that the multiple cathode-ray-tube beams would not move (upward). In this case "no vertical movement of the large number of cathode-ray-tube spots would be required. They would remain stationary, generating a single, horizontal, line pattern. The procedure would result in all of the (horizontal) one-dimensional zone plates in the radar record being superimposed to form one horizontal line. In the usual radar record, super-position of the one-dimensional zone plates occurs only for reflectors at the same range; in this situation, superposition occurs for all reflecting points. As in a true hologram, however (for which this superposition of zone plates of reflecting objects at all ranges also occurs), reconstruction of the original scene is not hampered by this superposition.[34]

It is interesting to compare the data acquisition of such a continuous-wave, stationary, synthetic-aperture (hologram) line-array radar with the data acquisition in a hologram. As we have already discussed (see Chapter 5, Reference 5), in the early days of laser holography it was difficult for many to understand how a single, two-dimensional, photographic plate, or a light beam having *two* degrees of freedom, could carry information about a *three*-dimensional object (described by three degrees of freedom). In the CW radar case described above, a *one-dimensional* line of superimposed zone plates carries full information about a *two-dimensional* area, just as the (synthetic) *single-line-array* of the aircraft captured the highly detailed infor-mation later to be reconstructed as the highly detailed photos of Figures 9.29–9.31.

In a 1973 work,[47] there is some discussion of the "laborious task" of the adjustment of the phase shifters to obtain the curved-wave-front focusing of the receiving array. We shall discuss a similar focusing situation in a near-field sonar in Chapter 10. One of the significant advantages of the hologram procedure in near-field sonar (or radar) is its ability to automatically provide focusing without the need for phasing procedures. The focusing is provided, as discussed earlier, through the zone-plate-forming action.

9.13. References

1. W. E. Kock and F. K. Harvey, Sound wave and microwave space patterns, *Bell Syst. Tech. J.*, **20**, 564 (1951).
2. J. R. Patty, Optical images of microwaves by wave reconstruction, NAVORD Rep. 6228, U.S. Naval Ordnance Lab., White Oak, Md., Jan. (1959).
3. R. P. Dooley, X-band holography, *Proc. IEEE* **53**, 1733 (1965).
4. D. E. Duffy, Optical reconstruction from microwave holograms, *J. Opt. Soc. Am.* **56**, 832 (1966).
5. G. Tricoles and E. L. Rope, Reconstruction of visible images from reduced-scale replicas of microwave holograms, *J. Opt. Soc. Am.* **57**, 97 (1967).
6. G. A. Deschamps, Some remarks on radio-frequency holography, *Proc. IEEE* **55**, 570 (1967).
7. N. H. Farhat and W. R. Guard, Millimeter wave holographic imaging of concealed weapons, *Proc. IEEE (Lett.)* **59** (9), 1383 (1971).

8. C. G. Goetz, Real-time holographic reconstruction by electro-optics modulation, *Appl. Phys. Lett.* **17**, 63 (1970).

9. W. E. Kock, Holographic computing in radar and ultrasonics, Invited paper presented at the Optical Computing Symposium, Darien, Conn., Apr. 12, (1972).

10. W. E. Kock, Real-time detection of metallic objects using liquid crystal microwave holograms, *Proc. IEEE* **60**, 1105 (1972).

11. W. E. Kock, J. L. Stone, J. E. Clark, and W. D. Friedle, Forward scatter of electromagnetic waves by spheres, *1958WESCON Conv. Rec.* **1**, 86 (1958); W. E. Kock, Related experiments with sound waves, *Proc. IRE* **47**, 1192 (1959).

12. D. Gabor, Presentation at 133rd AAAS Meeting, Washington, D.C., Dec. 30, 1966.

13. W. E. Kock, Scanning capabilities of metal lenses, Unpublished Bell Telephone Laboratories memorandum mm-46-160-14, Mar. 19, 1946.

14. W. E. Kock, Related experiments with sound waves and electromagnetic waves, *Proc. IRE* **47**, 1192–1201 (1959).

15. F. K. Harvey and W. E. Kock, Refracting sound waves, Unpublished Bell Telephone Laboratories memorandum mm-48-130-27, Aug. 25, 1948.

16. P. F. Checcacci, V. Russo, and A. M. Scheggi, Holographic antennas, *IEEE Trans. Antennas Propag.* **AP-18**, 811–813 (1970).

17. P. F. Checcacci, G. Papi, and V. Russo, A holographic VHF antenna, *IEEE Trans. Antennas Propag.* **AP-19**, 278–279 (1971).

18. C. F. Augustine, Field detector works in real time, *Electronics* **41**, 118–121 (1968).

19. C. F. Augustine and W. E. Kock, Microwave holograms using liquid crystal displays, *Proc. IEEE* **57** (3), 354–355 (1969).

20. C. F. Augustine, C. Deutsch, D. Fritzler, and E. Marom, Microwave holography using liquid crystal area detectors, *Proc. IEEE* **57** (7), 1333–1334 (1969).

21. M. King, Holographic processing of radar signals, *Laser Focus* **Aug.**, 43–45 (1969); M. Arm and M. King, Holographic storage of electric signals, *Appl. Opt.* **8** (7), 1413–1419 (1969).

22. E. Camatini, ed. *Optical and Acoustical Holography*, Plenum Press, New York, 1972, p. 11.

23. D. Gabor, The hologram, Friday evening discourse, Feb. 7, 1969, *Proc. R. Inst. G.B.* **43**, No. 200, 35–70 (1969).

24. L. J. Cutrona, E. N. Leith, C. J. Palermo, and L. J. Porcello, Optical data processing and filtering, *IRE Trans. Inf. Theory* **IT6** (3), 386–400 (1960).

25. L. J. Cutrona, E. N. Leith, L. J. Porcello, and W. E. Vivian, On the application of coherent optical processing techniques to synthetic-aperture radar, *Proc. IEEE* **54** (8), 1026–1032 (1966).

26. W. E. Kock, Acoustics and optics (introductory guest-editor paper), *Appl. Opt.* **8** (8), 1525–1530 (1969).

27. W. E. Kock, Holography and microwaves, paper presented at the U.S.–Japan Seminar on Holography, Tokyo, Japan, Oct. 5, 1967.

28. W. E. Kock, Side-looking radar, holography, and Doppler-free coherent radar, *Proc. IEEE* **56**, 238–239 (1968).

29. E. N. Leith and A. L. Ingalls, Synthetic antenna data processing by wavefront reconstruction, *Appl. Opt.* **7** (3), 539–544 (1968).

30. W. E. Kock, A method for extending the maximum range of synthetic aperture radar, *Proc. IEEE* **60** (11), 1459–1460 (1972).

31. W. E. Kock, Stationary coherent (hologram) radar and sonar, *Proc. IEEE* **56** (12), 2180 (1968).

32. W. M. Brown and L. J. Porcello, An introduction to synthetic aperture radar, *IEEE Spectrum* **6**, 52–62 (1969).
33. W. E. Kock, Holography, a new dimension for radar, *Electronics* **43** (21), 80–88 (1970).
34. W. E. Kock, Radar and microwave applications of holography, in: *Applications of Holography*, Plenum Press, New York, 1971 pp. 323–356.
35. E. N. Leith, Quasi-holographic techniques in the microwave region, *Proc. IEEE* **59** (9), 1305–1318 (1971).
36. W. E. Kock, Holographic computing in radar and ultrasonics, invited paper presented at the Optical Computing Symposium, Darien, Conn., Apr. 12, 1972.
37. G. L. Rogers, A new method of analyzing ionospheric movement records, *Nature* (*London*) **177**, 613–614 (1956).
38. G. L. Tyler, The bistatic, continuous-wave radar method for the study of planetary surfaces, *J. Geophys. Res.* **71** (6), 1559–1567 (1966).
39. J. J. Flaherty, K. R. Erikson, and V. M. Lund, Synthetic aperture ultrasonic imaging systems, U.S. patent 3,548,642, Dec. 22, 1970 (filed Mar. 2, 1967).
40. W. E. Kock, Nobel prize for physics: Gabor and holography, *Science* **174**, 674–675 (1971).
41. W. E. Kock, A hologram form of bistatic radar or sonar, *Proc. IEEE* **57** (1), 100 (1969).
42. W. E. Kock, Holographic techniques in continuous-wave bistatic radars, *Proc. IEEE* (*Lett*) **58** (11), 1863–1864 (1970).
43. W. E. Kock, Passive (cooperative) hologram radar, *Proc. IEEE* (*Lett*.) **58** (8), 1297 (1970).
44. W. E. Kock, Synthetic end-fire hologram radar, *Proc. IEEE* (*Lett*.) **58** (11), 1858–1859 (1970).
45. W. E. Kock, A holographic (synthetic aperture) method for increasing the gain of ground-to-air radars, *Proc. IEEE* (*Lett*.) **59** (3), 426–427 (1970).
46. L. C. Graham, Synthetic interferometer radar for topographic mapping, *Proc. IEEE* **62** (6), 763–768 (1974).
47. H. Ogura and K. Iizuka, Hologram matrix and its application to a novel radar, *Proc. IEEE* (*Lett*.) **61** (7), 1040–1041 (1973).
48. W. E. Kock, H. Ogura, and K. Iizuka, Comments on "hologram matrix and its application to a novel radar," *Proc. IEEE* (*Lett*.) **62** (6), 862–863 (1974); *Holosphere* **3**, Sept., 3 (1973).

10

ACOUSTIC HOLOGRAPHY APPLICATIONS

Since a hologram is a photographic record of the interference pattern generated between a set of waves of interest and a set of reference waves, it is obviously possible to make holograms of other forms of wave motion, provided the wave interference pattern can somehow be recorded. We saw earlier that visual presentations of wave progression for sound waves could indeed be recorded photographically (e.g., Figure 1.6) and that the process for doing this involved using a second set of waves as a reference set (Figure 1.5). We accordingly noted that figures such as Figure 1.7, 1.8, and 1.21, etc. can be regarded as acoustic holograms since they are recordings of interference patterns between waves of interest and a set of plane reference waves. The striations in these figures are acoustic fringes, exactly like the optical fringes formed when coherent light waves interfere. For those fringe patterns, no reconstruction process was performed, as the interest in those cases was in the fringe patterns themselves (the wave progression patterns). Nevertheless, to record acoustic fringe patterns for hologram use, the same technique is applicable, and acoustic holograms and their reconstructions have been made in this way in numerous laboratories.

Because the electronically introduced reference signal of Figure 1.5 corresponds to a plane-wave spatial reference whose wave fronts are parallel to the scanning plane of the microphone (as in Figure 1.6), an in-line hologram can be produced by orienting the scanning plane so as to be perpendicular to the waves of interest. This was done for Figure 10.1. The arrangement is thus similar to that of Figure 3.16, with the electronically injected reference

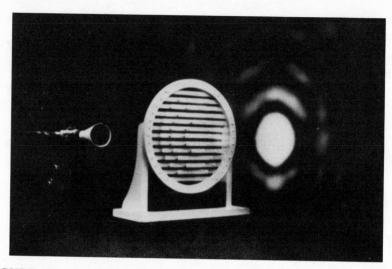

FIGURE 10.1. By causing the scanner of Figure 1.3 to scan a plane which is perpendicular to the direction of wave propagation, a cross section of the lobe structure of the acoustic radiator is portrayed (corresponding to the original in-line holograms of Gabor).

wave corresponding to the plane reference waves in the upper path of Figure 3.16, and the waves from the acoustic lens of Figure 10.1 corresponding to the spherical waves diverging from the right beam splitter of Figure 3.16. The hologram of Figure 10.1 thus resembles a zone-plate pattern, so that if the white rings of the figure were to be transformed into acoustically opaque structures, the result would resemble the zone plate of Figure 3.10 and would correspond to the somewhat similar acoustic zone plate of Lord Rayleigh.[1]

Because the directional properties of certain microphones (nonreversible ones, e.g., the carbon microphone, extensively used in telephone transmitters) cannot be determined by reciprocity (i.e., by using them as acoustic radiators), their directional patterns are obtained by using a scanning sound *source*. Thus, in Figure 1.5, the oscillator signal which energized the telephone receiver was fed instead to the scanning (radiating) unit, and the signal received by the telephone microphone was fed to the amplifier. Figure 10.2 was the result. It is of course quite similar to the pattern of Figure 1.6 since both units are small compared to a wavelength and therefore exhibit negligible directivity.

10.1. Early Developments

When the photograph of Figure 10.2 was first published,[2] the journal involved had learned of the experiments and had requested permission to publish a brief report on some of the results of the sounds portrayed. Numerous photographs were accordingly provided, with six being used, including Figure 10.2. The article's title was "Photographing Sound Waves" and the caption for Figure 10.2 was "Pattern of Sound Waves from a Telephone Handset." Observant readers noted that the carbon microphone was at the center of the circles and accordingly wrote to the journal, pointing out that sound does not issue from the carbon transmitter of a handset. As a result, a second article had to be printed, showing both Figures 1.6 and 10.2, and explaining the use of a scanning sound source in the case of the microphone pattern. Figure 10.2 is of interest in modern acoustic holography as it is the forerunner of the scanned-source procedure proposed 17 years later.[3,4]

In the article which included Figure 10.2, experiments were also described in which a succession of holograms such as the one in Figure 10.3 were recorded, with the relative phase of the reference wave changed slightly for each one, up to a value of 360°. A motion picture was then made, with these recorded in succession, "the result being a motion picture presentation of waves moving outward, . . . converging to a focus, and then diverging again."[2]

These experiments thus demonstrated several holography firsts, including the first use of a scanning source, an early movie using holograms (true

FIGURE 10.2. By causing the pickup microphone of the scanner of Figure 1.3 to *radiate* sound, and connecting the *microphone* output of the telephone handset to the scanning neon tube, the directional *receiving* properties of the carbon telephone microphone are made visible. This corresponds to the technique of scanning the sound *source* in making similar fringe patterns (acoustic holograms).

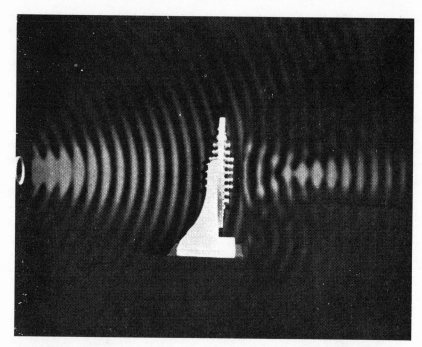

FIGURE 10.3. The fringes in this (static) pattern of sound waves being focused by an acoustic lens can be shifted slightly by modifying (slightly) the phase of the reference wave. Successive photographs, when repeated rapidly, provide the viewer with a motion picture of sound waves moving toward the lens and continuing on through the focal point.

hologram movies [5] were first reported much later) and, as noted in the excellent German text on holography, [6] the experiments involved the first recorded use of an off-axis reference wave in holography.

The pioneers in acoustic holography were F. L. Thurstone [7] and R. K. Mueller. [8] As we noted in Chapter 4, Mueller (of the Bendix Research Laboratories) first used, as an ultrasonic hologram surface, the liquid–air interface above the underwater, acoustically illuminated, "scene." In these experiments, a coherent reference wave was also directed at the liquid surface, and the surface then became the recording area for the hologram interference pattern (Figure 4.13). Because the surface of a liquid is a pressure-release surface, i.e., a surface which gives, or rises, at points where higher-than-average sound pressures exist, the acoustic interference pattern transformed the otherwise plane liquid surface into a surface having extremely minute, stationary, ripples on it. When this rippled surface was illuminated as shown in Figure 4.13 with coherent laser light, an image of the submerged object was reconstructed. We shall see that this liquid-surface technique has recently become important in the medical uses of holography. The procedure constitutes a real-time process which bypasses not only the photographic steps of optical holograms, but also the time-consuming scanning process used in recording acoustic interference patterns as in Figure 2.5. J. L. Kreutzer also employed acoustic holography at a rather early date[9] to demonstrate the ability of ultrasonic waves to portray, by holographic reconstruction, the screw threads in a solid structure (Figure 10.4).

In continued early experiments in acoustic holography, use of the piezoelectric, acoustic, equivalent of the image orthicon tube used in television systems, first proposed by Sokolov in 1937[10] and often referred to as an ultrasonic camera, was examined.[11,12] One arrangement is shown in Figure 10.5. Here the sound holography-generation technique uses a modified ultra-

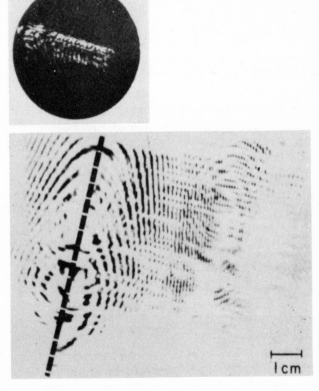

FIGURE 10.4. The reconstructed acoustic hologram image of a tapped hole (after Kreutzer).

sonic camera having a quartz crystal located at the front window of a cathode-ray tube. The quartz window is irradiated by an ultrasonic beam scattered from an object immersed in a water tank. The sound field incident on the quartz (piezoelectric) crystal induces a corresponding voltage distribution on it. This voltage distribution modulates the intensity of the secondary emission generated by the scanning electron beam. The resulting signal, which contains the information on the ultrasonic field scattered from the object, is then amplified and a reference wave is added electronically which stimulates an inclined acoustic reference beam.[12] This inclined beam is accomplished by causing the electronically injected reference wave to have a slightly different frequency from the illuminating sound source. Because the crystal is linearly scanned by the electron beam, the sampling points are detected at different times and the frequency difference between the illuminating and reference waves is translated into a phase difference, exactly equivalent to an inclined reference beam. (This process was later referred to as temporal reference holography.[13]) The interference "pattern" is then further amplified by the video amplifier and displayed on a television monitor. The images are reconstructed by photographing the hologram as displayed on the monitor and by illuminating the processed photographic plate with a laser beam.

The apparatus of Figure 10.5 was used to obtain the experimental results presented in Figures 10.6 and 10.7. A model of the letter C (lower left, Figure 10.6), was irradiated in transmission, and the hologram (right) was displayed on a television monitor. Photographing the television screen with a conventional camera provided the required transparency, which when properly irradiated with laser light generated the reconstructed image (Figure 10.7).

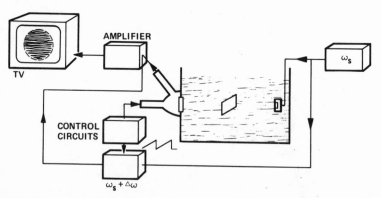

FIGURE 10.5. An ultrasonic camera used to generate acoustic holograms. The simulated electronic reference beam of a slightly different frequency is mixed with the detected signal derived from the sound scattered by object immersed in tank, which is irradiated from the right.

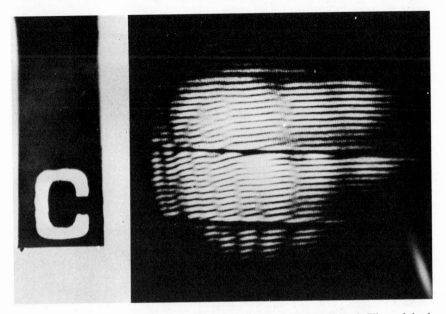

FIGURE 10.6. At the right is an acoustic hologram of the letter *C*. The original object is at the left.

10.2. Acoustic Synthetic-Aperture Systems

We turn now to synthetic-aperture hologram techniques in acoustic applications. One of the earlier proposals for such uses was described in a U.S. patent applied for in March 1967, entitled "Synthetic Aperture Ultrasonic Imaging Systems."[14] Two procedures were described, both aimed at medical ultrasonic applications for the "examination of the interior of living bodies." In the first (Figure 10.8), one transmit–receive transducer is moved (scanned) along a straight line; it is submerged in a liquid reservoir, with the liquid supported by, and in contact with, the body surface under examination. The second procedure utilizes a large number of fixed transducers positioned along the line of motion of the moving transducer of the other method (Figure 10.9), and energized in succession so as to simulate the moving transducer.

More recently, experiments using the ultrasonic synthetic-aperture procedure for true sonar use were described, with very excellent records resulting.[15] Figure 10.10 shows the arrangement of equipment used in these tests; the transmitter radiated 10-μsec pulses of 1-MHz sound waves. In these

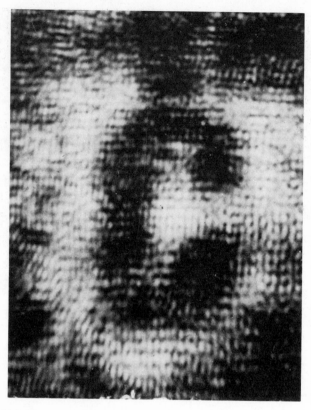

FIGURE 10.7. Reconstruction of the hologram of Figure 10.6. The outline of the letter *C* can be seen.

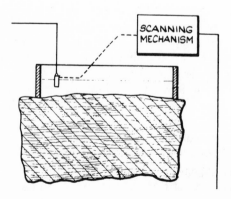

FIGURE 10.8. An ultrasonic synthetic-aperture system for medical applications.

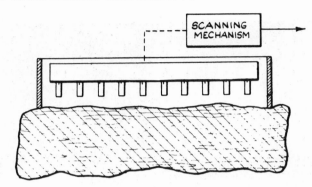

FIGURE 10.9. Energizing a series of transducers simulates the moving transducer of Figure 10.8.

experiments only the receiving transducer was moved. The round-trip distance to each of the three reflecting objects from transmitter and receiver was made sufficiently different so as to permit the one-dimensional zone plates generated by the reflections from each of the three objects to be adequately separated in the holographic record. For these tests, the reference signal was supplied electronically from the stabilized oscillator directly to the receiver, which also supplied energy (amplified) to the transmitting transducer (as in Figure 1.5). In Figure 10.11, the three, one-dimensional, zone plates (one for each

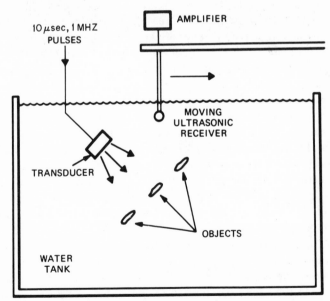

FIGURE 10.10. Arrangement for synthetic-aperture (hologram) sonar experiments (after Pekau and Diehl).

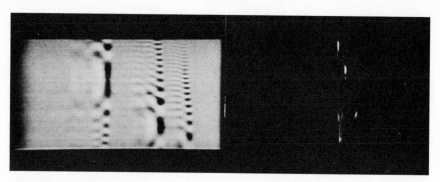

FIGURE 10.11. Zone-plates generated by the equipment of Figure 10.10 (left). The three top dots at the right are the reconstructions of the objects of Figure 10.10 (after Pekau and Diehl).

reflecting object) are seen, all adequately separated to permit reconstruction, from the zone plates, of the three individual objects.

The similarity of these zone plates to the radar zone plate of Figure 9.27 is evident. It is seen that the left-hand zone plate has its central portion slightly above center of the photograph, whereas the middle zone plate has its central portion slightly below, and the right-hand one has its central portion below the bottom of the photograph. The right-hand portion of Figure 10.11 shows the reconstruction (by laser light) of the three reflecting objects (the two light areas at the lower left are artifacts). One of the two co-investigators in this work (D. F. Pekau) had earlier been a colleague of R. K. Mueller at the Bendix Research Laboratories. We shall discuss a more recent [16] application of the acoustic synthetic-aperture procedure in Chapter 13 (medical applications).

One of the limiting factors in the synthetic-aperture process results from the need to delineate accurately the finer fringes of the zone plates. As the platform carrying the transducer moves, the distance to the reflecting object changes, and a round-trip change of one half wavelength causes the interference between the fixed reference wave and the varying echo wave to change from a constructive interference case to a destructive one. The former corresponds to the black portions of the zone plates of Figure 10.11, and the latter to the white, or blank, portions. It is obvious that if too few pulses return during the period when a "fringe" (adjacent white and dark areas) is being generated, the zone plate cannot be delineated properly.

This need to send out closely spaced pulses causes the maximum useful (unambiguous) range to be limited since the time interval between pulses corresponds to the round-trip time to the most distant targets. While this is somewhat annoying in the radar case, it is far more serious in the sonar case because of the much lower velocity of sound waves. Thus, in a recently published report of a National Academy of Sciences Summer Study, it was noted that for a ship traveling at a speed of 6 knots and carrying a 1-kHz synthetic-aperture sonar, the maximum unambiguous range would be only 1.33 nautical miles (see Chapter 9, Reference 30).

As noted earlier, a procedure was recently described which overcomes this range problem (see Chapter 9, Reference 30). The concept is illustrated in Figure 10.12. The usual single transmit–receive transducer case, with its need to transmit pulses whenever the sonar moves a distance equal to one-half the aperture dimension D (see Chapter 9, Reference 32), is shown in the upper left of Figure 10.12. In the new procedure, shown at the bottom left, a receive-only transducer is added ahead of the transmit–receive one. It is evident that the signal received by this second unit for pulse 1 is the same as that received for pulse 2 in the usual (upper left) case, since for that case, the outward path is shorter and the receive path is longer. The pulse repetition

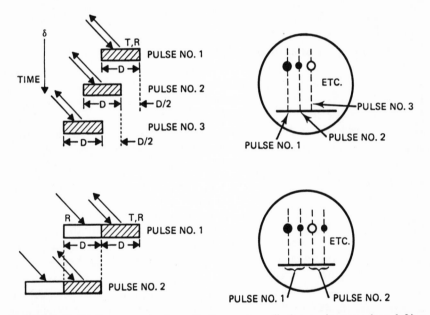

FIGURE 10.12. In a standard synthetic-aperture (hologram) sonar (top left) a new pulse must be transmitted each time the moving platform advances one-half the aperture dimension D. The top right portion indicates how a target generates, with successive pulse echoes, a horizontal, one-dimensional zone plate on the cathode-ray tube of Figure 9.22. The heavy black dot corresponds to a constructive interference situation, and the white circle to a destructive one. The bottom portions of the figure show how the pulse spacing (and maximum range) can be doubled through the addition of a receive-only transducer whose signal is fed to a second cathode-ray-tube beam moving up with the original one.

rate can therefore be halved. If instead of one, three receive-only transducers are added, the pulse repetition rate can be reduced by a factor of four (with a consequent increase in the maximum unambiguous range by that same factor). If seven receive units are added the increase becomes eight, etc. The cathode-ray-tube patterns (comparable to those of Figure 9.22) for the two cases are shown at the right; for the second case, the tube must be equipped with two adjacent, upward-moving beams, with the second one amplitude-modulated by the receive-only transducer signal (properly combined with the reference wave, i.e., synchronously demodulated). This recent development could materially extend the usefulness of holographic (synthetic-aperture) sonars.

We saw earlier that the synthetic gain associated with a synthetic end-fire radar or sonar can cause an air-borne, forward-looking (synthetic), linear array to be effectively generated (see Chapter 9, Reference 44). For objects in the direct path of the moving vehicle, the hologram process (whereby the coherent reference wave is provided by the process of synchronous detection) here generates one-dimensional gratings rather than zone plates. Such *exactly* forward-looking systems are not to be confused with air-borne radars which gain information on the ground terrain which is forward of the aircraft. Because these gratings possess an extremely small fringe spacing, the ambiguity constraint here is quite severe. Thus, in one suggested arrangement for the possible detection of clear-air turbulence (see Chapter 9, Reference 44), a frequency of less than 250 MHz was required (because of the ambiguity constraint) to achieve a maximum detection range of 93 miles. When the technique just described for extending the maximum range is applied to the end-fire synthetic-gain radar, one additional receive-only antenna placed ahead of the standard transmit–receive antenna would permit (for the case described) the maximum range (93 miles) to be doubled, or the frequency (250 MHz) to be doubled. With three receive-only antennas in that case, the range or frequency could be quadrupled, with seven, both could be increased by a factor of eight, etc. Sonar systems acquire similar benefits.

In the synthetic-aperture-related stationary case, the hologram procedures provide a small, wide-angle, radar or sonar with a sizeable directivity gain against targets moving past it with uniform velocity along a straight line (see Chapter 9, Reference 45). Here again, the use of additional receive-only antennas with their signals properly processed can (1) materially extend the maximum range capability of such stationary systems (through the resulting satisfactory delineation of the finer zone-plate fringes with a lower pulse repetition rate), (2) permit higher-frequency waves to be used, or (3) enable the system to be effective against much-higher-velocity targets. For stationary synthetic-gain systems designed for targets coming directly toward them, in which end-fire gain is generated through the use of hologram gratings rather than zone plates, additional receivers again can (1) increase the maximum range capability, (2) permit the use of higher frequencies, or (3) enable the equipment to detect higher-speed targets. Any or all of these effects could be advantageous in the over-the-horizon radar application against cruise missiles (see Chapter 9, Reference 45).

Additional receivers are also useful in CW bistatic systems when placed at the receiver location and arranged along a line perpendicular to the line joining transmitter and receiver. One additional receiver can indicate the direction in which a target is moving when it crosses the line joining the transmitter and receiver (this information is not available in the usual Doppler-detecting, bistatic system). This is given in the two-receiver case by the time difference indicated at the point of zero Doppler of each receiver. A further advantage is attainable through the use of many additional receivers. By separately recording all receiver signals (photographically, as zone plates) and then, in the reconstruction process, repeatedly shifting all records from a wide (but uniform) individual separation in time to a narrow separation, and noting any point having a large signal-to-noise improvement, the position of this focused signal spot provides the velocity of crossing of the target, and also, at that spot, a significant signal-to-noise improvement, as generated by the many exactly superimposed zone plates (as many as the number of receivers).

Another example for which the pulsed, synthetic-aperture, multireceiver concept may be useful involves side-looking sonar applications. Presently, it appears that such sonars have not yet exploited hologram concepts. Figure 10.13 shows a very interesting but rather complicated side-looking sonar

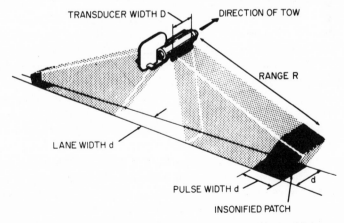

FIGURE 10.13. A deep-submergence, variable-focus, side-looking sonar based on a concept of V. C. Anderson and F. N. Spiess (courtesy of the Marine Physical Laboratory of Scripps Oceanographic Institution).

designed and used by the Marine Physical Laboratory at the Scripps Oceanographic Institution, with support from the Navy Deep Submergence Project Office.[17]

The sonar operates at 400-kHz, using a 3 m linear array of 29 elements for providing a picture of the sea floor. The ranges of interest are thus quite short (9–230 m), all being within the near field of the 3-m array. To cover this range of distances, a 24-step electronic sequence is performed following each pulse transmission in which the required delays for each 10-m range increase are provided so as to properly focus the array inward at each near-field point in the object space. These delays, all different for each of the 29 elements and for each of the 24 range steps, must be incorporated very rapidly so as to have the array properly focused to accept the multitude of returning sonar echoes. In spite of this rather awesome electronic complexity, the sonar has been assembled and is now providing excellent "pictures" of very deep ocean-bottom areas. The side-looking sonar geometry is shown in Figure 10.13.

The transducer is flown off the bottom and echo ranges to either side. By operating in the near-field region of the transducer, the system scans a constant-width lane on the bottom out to a slant range R. If the pulse length is made equal to the lane width, the insonified patch which contributes to the reverberation at any one instant of time has an area of d^2. In the simple system, the time required to advance the transducer one lane width along the direction of travel must be no less than the two-way travel time of the echo-ranging pulse to the limiting range R. Thus, the basic area search for such a system is given by the product of the lane width and the velocity of propagation in the medium.

Inasmuch as one of the real advantages of pulsed hologram (synthetic-aperture) sonar is its ability to automatically provide near-field focusing without either focusing or beam-forming delay elements being required, it would appear that hologram concepts would be quite useful here. As we noted earlier, this near-field focusing ability in the radar case was very difficult for many to understand until the zone-plate-forming action was recognized.

Recent experiments[18] involving the long-range transmission of fairly low-frequency (367 Hz) sound waves, have shown that such sound waves exhibit fairly high coherence over a rather long path (700 nautical miles) and over a reasonably long period of time (105 sec). The conclusions drawn from these experiments suggest that hologram techniques might be useful in underwater applications at these lower frequencies. Thus, the statement was made[18] that the results were "taken to be evidence that the sound velocity structure is frozen in the volume of the ocean, or at least partially so, over this period" (105 sec). These experiments were conducted in the Atlantic Ocean using a bottom-mounted source located off the island of Eleuthera in the Bahamas and a receiving array off Bermuda.

In the holographic utilization of the high coherence of low-frequency sound in the sea, the stationary forms of hologram sonar, as described in Section 9.7 would appear to be the most likely candidates. Thus, use might be made of line-array receivers (or the partial line array of Figure 9.34), in monostatic systems or in bistatic systems. The latter could use a transmitter radiating to a receiving *array* (see Chapter 9, Reference 41) for detecting stationary targets, or to a single receiving transducer (see Chapter 9, Reference 42) whereby the received zone-plate signals (the V curves of Figure 9.39) as generated by a moving target (Figure 9.40) would be processed so as to utilize the gain inherent in such signals. Stationary, monostatic hologram systems (Figure 9.42) could also be of interest, again against moving targets, in either the side-looking arrangement or the end-fire one.

Figure 10.14 shows an early low-frequency transmitting array (designed by the Bendix Research Laboratories, and also used in acoustic transmission experiments between Bermuda and Eleuthera) about to be lowered into the Atlantic Ocean from a ship located just south of Bermuda. Pulses of low-frequency (250-Hz) sound transmitted from this ship-supported unit were received both at Bermuda and at Eleuthera (Figure 10.15). Those received at the nearby Bermuda receiving point (top) exhibited reverberation effects which invariably accompany a strong energy pulse. At the much more distant

FIGURE 10.14. An experimental, deep-water, low-frequency, acoustic sound source (Bendix Research Laboratories).

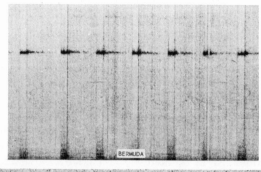

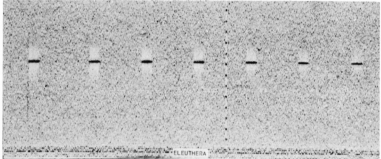

FIGURE 10.15. Transmissions (top) from the sound source of Figure 10.14 as received 700 miles away (bottom).

Eleuthera receiver, the local Bermuda reverberation artifact was too weak to be received and the pulses were seen to be very clear and sharp (bottom). These pulses were 30 sec in length, and the reverberation in Bermuda was seen to last for several seconds.[19] Long-range coherence measurements were also made by cross correlating the transmitted and Eleuthera-received pulses, and a quite definite indication of coherence was observed, providing further support today for possible holography uses. Figure 10.16 shows the transmitting array of Figure 10.14 mounted on a supporting structure as it was later being lowered to the bottom off Eleuthera to act as a bottom-mounted transmitter for further transmission tests from Eleuthera to Bermuda. Results of such tests, using this bottomed unit, were later reported by Bell Laboratories scientists.[20]

10.3. Underwater Applications

Because sound waves are the only form of waves which are able to propagate to any significant distance in sea water (or fresh water), the use of acoustic devices for detecting underwater objects (such as submarines) has been explored and used for many decades; in fact sound-wave echo-location systems (later referred to as sonar systems) were used during World War II. In more recent times, sonar design has developed into a highly sophisticated technology, so that some who are involved in the design of modern sonars often tend to dismiss the thought of applying holography to sonar because of their conviction that the later techniques are surely already being employed. Also, because the concept of phase has been associated with steered-beam sonars for quite some time, the stress often put on the phase-recording ability of a hologram has again led some sonar experts to feel that the technique provides nothing new.

Now it is true that as yet acoustic holography has not been able to make much use of three-dimensional properties which are so striking in optical holograms. We noted in connection with Figure 5.15 that even a much

FIGURE 10.16. The transducer array of Figure 10.14 arranged for deep-bottom mounting.

smaller portion of a hologram (or the zone plate of Figure 5.15) can be used to reconstruct the original scene. However, it is evident from that figure that the central portion is like a smaller window, so that the three-dimensional parallax effect is less useful for it. Thus, in Figure 5.15, the viewer is required, for the area, $A'B'C'D'$, to position himself more carefully in order to see the source through this new, smaller window. Also, a small window reduces appreciably the realism he would obtain in a *full* hologram through the parallax effect. With the smaller window, the viewer can still move his head about and see all of the objects in the scene, but the window area he is using does not move with him, and he merely sees small, individual, almost two-dimensional views. As in the case of a true glass window, the viewer's ability to see around objects in the foreground is limited when the window size is reduced to a significantly smaller area.

Because of the much longer wavelengths of sound waves as compared to light waves, even fairly large acoustic holograms act like small windows, so that when they are optically reconstructed very little parallax or three-dimensional effects are available. Another way of stating the above is to say that the average acoustic hologram portrays very little of the near (Fresnel) field. (An exception to this would be a hologram made from the receiver outputs of the array of Figure 10.13, which is intended for near-field use.) The near field is usually defined as extending out to a distance $a^2/2\lambda$, where a is the aperture dimension and λ the wavelength. Thus, the near field of a 50-wavelength aperture would extend out to a distance equal to 1250 wavelengths. If an ultrasonic frequency of 25 KHz is used ($\lambda = 0.2$ ft), the aperture would be 10 ft and the near field would stop at 250 ft.

Because many of the advantages of holography and synthetic-aperture systems relate to the ability to provide detailed information concerning objects located in the *near* field, some are tempted to dismiss consideration of the far-field performance of hologram sonar as unimportant, since ordinary sonar performs this task well. On the other hand, hologram techniques often provide simpler ways of achieving the performance requirements of the usual, far-field, sonars.

Consider, for example, a multielement, square-array sonar, having 100 elements on a side, and hence 100×100 or 10,000 total elements. If the elements are positioned $\lambda/2$ apart, the aperture dimension is 50λ on a side. The near field, as noted above, would end for this case at 1250λ. We now suppose further that the array itself is used only for receiving, and that the transmitted signal, radiated by another, high-power transducer, illuminates a pyramidal volume of sizeable extent. The array is to acquire, to the best of its ability, information regarding reflecting objects located in the far field and within the radiation pyramid of the transmitting transducer.

One way of accomplishing the desired function of the array would be to establish thousands of "preformed" beams, whereby, for any single beam direction, phase shifts would be provided as necessary for each of the 10,000 array elements, so that all elements would contribute properly to the receiving beam for that direction. Because the array is 50 wavelengths on a side, the width of each of the preformed beams will be approximately $1°$ ($51\lambda/a$), so if a $90°$ pyramid of coverage ($\pm 45°$) is desired, approximately 90×90 or 8100 preformed beams would be required. Whether the required phase shifts (they total, for 10,000 elements and 8100 beams, approximately 80,000,000) are provided individually (by analog methods) or by digital-computer techniques, the task is obviously not a simple one.

Consider now providing these same far-field beams by holographic sonar procedures. The coherent transmitted signal would be provided by a stable oscillator, and a small amount of this oscillator signal would be used as a reference signal. It would be supplied continuously to each element of the receiving array (possibly by direct line to the element), with the phase adjusted so as to simulate a plane wave of this frequency impinging (preferably at some angle) on the entire array. Reflecting objects located within the cone of the transmitter will reflect some of the transmitted signal back to the array, thereby generating, in combination with the reference signal, a holographic interference pattern which is "sampled" by all the elements. The resultant values at all elements are then recorded photographically, and, after development of the photographic record, the far-field reflecting objects would be reconstructed by coherent light. (As noted, we shall see later that real-time reconstruction methods are now under development.)

At first glance it would appear that the first procedure which combines all 10,000 elements into a preformed beam, thereby achieving an extremely high directivity gain [40 decibels (dB), or so], *must* be superior to the procedure in which each single, isolated unit of the 10,000 elements is made to affect, by itself, the exposure of the photographic plate at that point.

In the 10,000-element preformed beam a reflected target signal arrives in phase at all elemental receivers (after proper phase shifts are taken into account) and the summed output is therefore very large; for the same elements, the sea noise, on the other hand, coming from all directions, adds in a *random* way, and the noise-signal buildup is thus far less. In the hologram case it would appear as though each elemental receiver, with its own amplifier, treats noise and signal equally, thereby losing the array gain of the preformed beam. Actually, the coherence of the signals generating the interference pattern on the array and the similar coherence of the laser reconstructing beam, causes each recorded point to combine coherently (during the reconstruction process) with all the rest of the points, thus providing, holographically, a comparable array gain for each of the 8100 directions.

An alternative standard design for the 100×100-element receiving array would employ *one* receiving beam, made steerable to all of the 8100 directions through the use of a much smaller number of *variable* phase-shift mechanisms. This technique however loses the important time-integration advantages accrued through individual storage of the signals in each of the outputs of the 8100 preformed beams, or equivalently, the storage of each of the receiver-element outputs in the hologram version, since the scanning beam looks in any one direction for only 1/8100th of the time.

This consideration also brings out an interesting point regarding the ability of the medium to maintain the extraordinary coherence which is usually assumed to be necessary in hologram sonar. Thus, it is known that the stable oscillator signal of acoustic hologram systems can have extremely long coherence lengths, *in a perfect medium*. But because sea water introduces velocity variations due to currents, thermal fluctuations, etc., many assume that a holographic sonar will automatically suffer far more than an ordinary sonar from these effects. However, for a large array (such as the 100 × 100 wavelength, one just discussed) the performance of the standard form of sonar will also be degraded by velocity variations. A reflected signal will only arrive at the array with a perfectly flat wave front if the medium is perfect, and hence any changes in *spatial* coherence caused by the medium will degrade the performance of a standard sonar array designed to receive flat wave fronts. Also, the phasing methods used in forming or steering sonar array beams are often highly frequency sensitive, so that an impairment, by the medium, of the frequency coherence of the signal will also affect performance. It is because of these effects that beam output integration is useful, since the magnitude of the variations in spatial and temporal coherence as caused by the medium is variable with time, and a steered beam might be aimed in the target direction exactly at a time when the coherence impairment is a maximum, with the result that the received target signal may at that instant be undetectable. This need for beam-output integration tends to force the designers to the very complicated, preformed-beam version, for which the hologram form offers such an attractive, simpler alternative.

Recently, in a program funded by the U.S. Navy Office of Naval Research, a holographic imaging system was constructed that included a 400-element hydrophone array (Figures 10.17–10.19) with all the necessary electronics, plus a photographic recorder, permitting the successful reconstruction of acoustic targets. This is now being followed by the design and construction of a large-scale, deep-operating, holographic imaging system using a 9216-element hydrophone array. It is a square array (96 elements on a side), and, based on the successful results of the earlier 400-element array, it is expected to provide a resolution of 0.00487 rad (between 0.3° and 0.4°). One of the most important components of this system is a real-time reconstructor tube (see Chapter 9, Reference 8) shown in Figures 9.6 and 9.7.

FIGURE 10.17. A 400-element receiver array for underwater acoustic holography.

In this latest sonar system, an ultrasonic signal is sent out by the transmitter and reflected by objects (which can in general be *moving*). The reflected waves are detected by the large receiving array, and the resulting electrical signals are converted into holographic information by an electronically injected reference wave. The reconstructor tube converts the holographic signals into real-time visible images of the object scene, and these are then displayed for the operator.

The reconstructor tube, as we noted in the Chapter 9, consists of a DKDP crystal, an off-axis scanning electron gun, and associated optics. In operation, the scanning electron beam is modulated with the holographic information so that a hologram is written on the crystal in the form of a positive charge

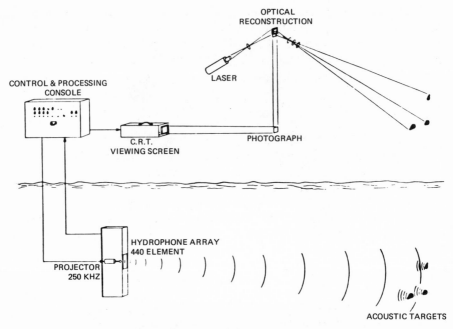

FIGURE 10.18. Elements of a holographic sonar system.

pattern. The electric field within the DKDP varies over the crystal according to the holographic signal, thus modulating its refractive index by the electro-optic effect, so that coherent light transmitted *through* the DKDP crystal becomes modulated with (that is, it reconstructs) the holographic information. The hologram is periodically erased by flooding the crystal with electrons with an appropriate potential on a nearby grid. Moving objects generate varying holograms, and the real-time reconstruction follows the variation which occurred in the object illuminated. The real-time capabilities are now of the order of 16 frames/sec. Figure 10.20 shows a hologram, and Figure 10.21 shows the reconstruction of it as accomplished by this tube. As an example of a possible configuration for using this underwater viewing system, Figure 10.22 shows an artist's concept of an experimental search vehicle.

10.4. Acoustic Kinoforms

Because many electronic devices such as cathode-ray tubes can generate visible light patterns which can be recorded photographically, computer scientists recognized rather early that cathode-ray-tube signals, originating from a properly programed computer, could become, when photographically recorded, computer-generated holograms. Numerous such holograms and their reconstructions have been reported,[21-23] with A. W. Lohmann probably being recognized as the pioneer in this field. More recently, a very interesting form of computer-generated hologram called the kinoform has been described,[24,25] interesting because it accomplishes what normal (planar) holograms cannot do: it eliminates both the straight-through (zero-order) component and one of the two first-order diffracted hologram components (either the virtual or the real). Although it thus constitutes one of the most efficient holograms, its application possibilities are not as yet fully determined. We will discuss it here because it has some interesting counterparts and possible applications in acoustics.

The kinoform, in a general sense, is a computer-generated wave-front-reconstruction device, which, like the hologram, provides a display of a three-dimensional image. When illuminated, however, it yields only a single diffraction order, so that, ideally, all the incident light is used to reconstruct one image. The kinoform may be thought of as a complex lens which

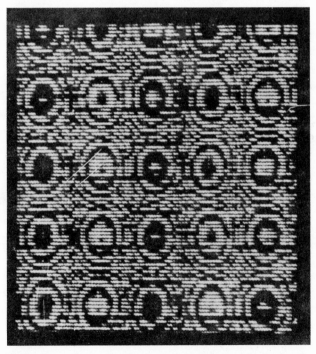

FIGURE 10.19. Zone-plate records generated by groups of transducers in the array of Figure 10.17.

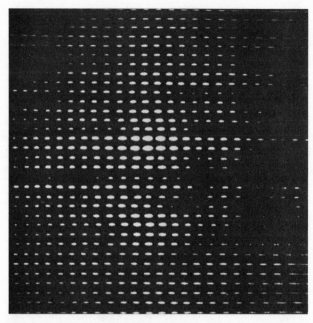

FIGURE 10.20. Acoustic hologram of the letter E taken by means of a scanning transducer.

transforms the reference wave incident upon it into the wave front needed to form the desired image. Although first conceived as an optical focusing device, the kinoform can be used as a focusing element for any physical wave form, including sound waves. The useful information in a kinoform, is, like that in a hologram, a coded description of the wave front of light scattered from a particular object of interest. Upon illumination with a reconstruction beam, the kinoform provides the display of an image.

The kinoform concept can perhaps be best explained by a comparison of Figures 3.2 and 3.6. In Figure 3.2 the slit variety of a grating and the photographic variety are sketched; both diffract several orders, but the latter diffracts only the zero order and the the two first orders. In Figure 3.6, a periodic array of prisms is shown; this too can be considered as a form of grating, but because the diffracting elements are dielectric prisms rather than slits or sinusoids, only one diffracted order results (even the zero-order, straight-through, waves are eliminated). For this latter form of grating, the prisms must be designed rather carefully so as to cause all elemental, prism-refracted waves to combine properly with all other such elemental waves. Now it has been known for some time [28] that if the positions where the prisms are stepped back are points of 2π-rad phase delay, then all of the elemental

FIGURE 10.21. Real-time reconstruction of Figure 10.20 with the DKDP light modulator of Figure 9.6.

diffracted waves add. This is illustrated in Figure 3.6 by the parallel-wave-front lines being spaced one wavelength apart.

This same relation holds when the (refracting) grating shown is altered so as to become a lens. Thus, whereas a circular zone plate (such as the photographic one illustrated in Figure 3.13) generates a zero-order wave and two sets of first-order diffracted waves, a zoned (stepped) *lens*, patterned

FIGURE 10.22. Underwater viewing using acoustic holography.

after the stepped prism of Figure 3.6, generates only one wave set. If the lens is designed to be a converging (focusing) lens, the waves converge up on a (real) focal point; if it is a diverging lens, the waves diffracted (and refracted) by it emerge as waves *diverging* from a (virtual) source.

A microwave lens, fabricated in 1944, for use in a parallel-plate antenna[26] is shown in Figure 10.23. The maximum step thickness corresponds to a phase change in the dielectric of 2π rad as outlined above, and because the dielectric has a refractive index which is greater than 1 the lens shown produces converging waves. It is of interest to compare this lens with Figure 19.10 of Reference 27, which was used there to explain kinoforms. (The lens of Figure 3.7 is also a microwave kinoform lens; the refractive material has, however, an index less than 1 (it is a parallel-plate waveguide design) so for the lens to be converging (convex) the contours must be concave.[28] Again the steps are placed at those points where the thickness reaches a value equal to a wavelength in the refractive material (2π rad). The similarity of this microwave lens to a kinoform was recently noted.[29]

In the more general form of optical kinoform, a complicated object is reconstructed. However, one must also be able to express the hologram for this object analytically so that a computer can indeed generate the light pattern (for example, on a cathode-ray-tube). Again, as was discussed in connection with Figure 4.3, each point of the reconstruction can be looked upon as being generated from a simple kinoform lens (a circular form of the one in Figure 10.23 but drastically reduced in size, of course, so as to be functional at optical wavelengths). The superposition of all such lenses causes it to become the "complex lens" which then transforms the reference wave incident upon it into the wave front needed to form the desired image. In an optical kino-

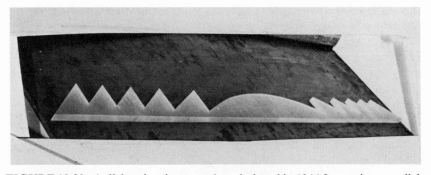

FIGURE 10.23. A dielectric microwave lens designed in 1944 for use in a parallel-plate antenna. Because it is stepped at one-wavelength thickness, it can be considered a kinoform lens.

FIGURE 10.24. An experimental zoned (stepped) parabolic reflector designed as a wide-angle scanning microwave antenna (Bell Telephone Laboratories).

form, the varying thickness and the steps at 2π rad of Figures 3.8 and 10.23 are obtained by bleaching the film in a very careful way, so that the hologram becomes a combination diffracting and refracting device (like the prism array of Figure 3.8) instead of a purely diffracting one.

Kinoform lenses have been considered for use in underwater acoustic systems because they can be thin and lightweight, thus minimizing attenuation and mounting considerations. Recently an ultrasonic technique was described in which a kinoform mirror was used as a focusing element.[30] The mirror was a reflective variant of the simple kinoform lens and was constructed so as to transform an incident plane wave into a converging spherical one. The limitation, in optics, which exists for kinoform lenses or reflectors, namely that they must be used with nearly monochromatic radiation, is not so serious in ultrasonic applications, where single-frequency sound waves are often employed.

The kinoform mirror is, in effect, a stepped paraboloidal reflector, a form investigated in the microwave field during World War II (Figure 10.24). Figure 10.25 shows the usual form of paraboloidal reflector, with its parabolic cross section, and Figure 10.26 shows the recently described stepped paraboloidal reflector.[30] The cross section of the lens of Figure 3.7, on the other hand, is that of a stepped ellipse. Figure 10.27 shows how the 2λ-rad steps must be formed to achieve proper phase correction. The top equation is that of the cross section of the basic ellipsoid of revolution (corresponding to the paraboloid of revolution of the reflector of Figure 10.26). When a lens

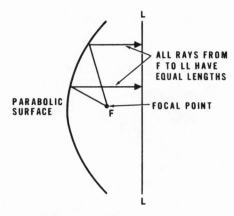

FIGURE 10.25. A parabola used as a focusing reflector.

FIGURE 10.26. A flat, two-dimensional, stepped parabolic reflector, i.e., a kinoform reflector (IBM, Reference 24).

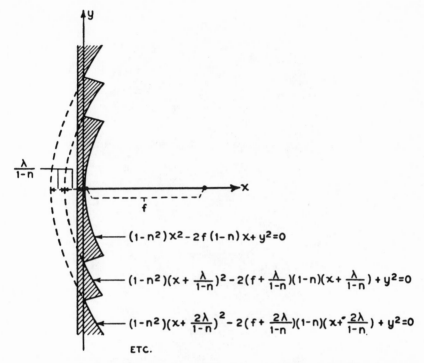

FIGURE 10.27. Step design for the elliptical contours of a waveguide microwave lens (a microwave kinoform).

thickness corresponding to a phase shift of 2π rad or one wavelength is reached, the lens is stepped back and a new elliptical contour is followed corresponding to a focal-length increase of one wavelength in the refracting medium (for a reflector the increase is a half wavelength). The process is then continued. The stepped shaped and curved sections are required to ensure that the wave energy is refracted into a single diffraction order with maximum efficiency. Obtaining such an intricate relief at optical frequencies (in hologram kinoforms or kinoform lenses) through the highly controlled shrinkage of the emulsion poses certain difficulties. However, at the wavelengths considered useful for ultrasonic applications, machining procedures are considered adequate in fabricating the reflector of Figure 10.26. It was designed as an f/l kinoform mirror for use with 5-MHz ultrasound in water ($\lambda =$ 0.29 mm). The mirror is 50 mm in diameter and was made on a lathe. As noted earlier, the steps here are at $\lambda/2$-thickness points because it is a reflector rather than a lens. A comparable, circular, microwave lens, with index of refraction less than 1, is shown in Figure 10.28. As we noted in connection

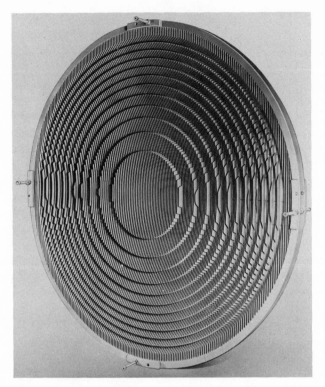

FIGURE 10.28. A circular microwave kinoform lens. The resemblance to the kinoform reflector of Figure 10.26 is interesting.

with Figure 9.8, both the flat reflector of Figure 10.26 and the flat lens of Figure 10.28 could have been made to exhibit less aberration if they had been made spherical instead of flat. The reason for which the microwave parabolic reflector enclosed in the World War II antenna of Figure 10.24 was stepped was to permit it to be given a circular shape and hence a wider field of view (less coma aberration).

Acoustic kinoforms (variable-thickness reflectors or refractors) might be useful in ultrasonic, nondestructive testing procedures. Thus, for example, a kinoform could be designed to convert a plane acoustic wave into a multiplicity of narrow beams, all aimed in predetermined directions, so that when used for testing a particular structure which is opaque to light waves but transparent to sound waves and which should have a number of properly positioned hollow areas (voids), reflections caused by the voids could be ascertained by sensors placed in the paths of the acoustic kinoform sound beams.

10.5. Holographic Pulse Compression

In 1970, a method of using holographic methods in pulse-compression sonar was described.[31] Pulse-compression techniques, as used in both radar and sonar, convert long, high-power transmitted pulses into short, high-resolution, received pulses. Most transmitting transducers have a definite overload point, which places a limit on the signal amplitude they can handle. Hence, to put more power into the outgoing pulse, the only alternative is to increase the pulse length. But long pulses, returning from two targets at slightly different ranges, will overlap, preventing the system from resolving the two targets. In Figure 10.29, two different-length, equal-amplitude sonar pulses are shown, and it is evident that the longer, more powerful pulse cannot resolve two targets spaced apart, say, at a distance corresponding to twice the length of the short pulse. Through the use of pulse compression, this difficulty is overcome.

In most of the presently used pulse-compression systems, the frequency is varied during the duration of the pulse,[32] and the received pulse is compressed in length by means of a matched, dispersive filter. Because of the rising frequency variation imparted to the pulse, the name "chirp" has been associated with such systems. The dispersive filter causes signals of different frequencies to travel at different velocities, thereby permitting the more rapidly traveling portion of the pulse to catch up with the slower, earlier-generated portion, resulting in a shortening of the pulse length.

It is of interest to note that many of the early concepts involved in the chirp process were considered almost simultaneously at the New York, Murray Hill, and Holmdel Laboratories of the Bell Telephone Laboratories.

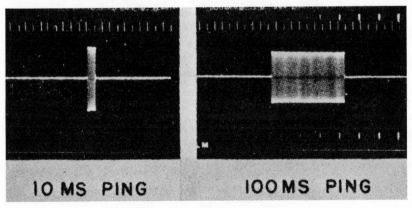

FIGURE 10.29. Sonar pings of different lengths.

This author approached the problem (at the Holmdel Laboratory) through "velocity modulation," the process which is the basis of the velocity-modulated microwave tube, the klystron. In this tube, a stream of electrons has its velocity changed so as to permit those electrons emitted later (from a cathode) to catch up with those emitted earlier. The tube contains a "buncher" to which a varying voltage is applied to cause electrons leaving the buncher at a later time to overtake those leaving earlier. Figure 10.30 shows how a particular velocity variation can cause electrons to arrive simultaneously at repeated points, permitting the "bunched" electrons to repeatedly excite a microwave cavity resonator and thus generate microwaves.

The author's variation of this process, which he referred to as "wave bunching," involved waves propagating in a waveguide. In the development of the waveguide microwave lens [28] of Figure 3.7 it was recognized that the wave velocity within a waveguide was dependent upon wave frequency, so the thought put forward was that a *wave velocity* modulation could be brought about through a *frequency* variation, so as to cause waves introduced at a later time to overtake waves generated earlier having a lower frequency.

This waveguide effect was demonstrated much later (see Chapter 9, Reference 14) for sound waves (Figure 10.31). Figure 10.31(A) shows two superimposed acoustic pulses of different frequencies entering an acoustic

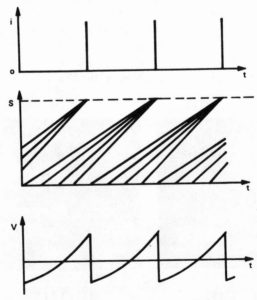

FIGURE 10.30. Velocity variations (bottom) imparted to electrons (middle) cause them to arrive together (top).

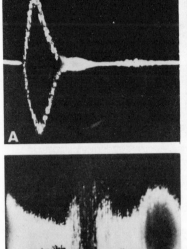

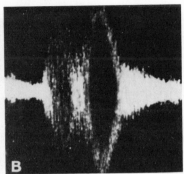

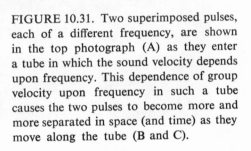

FIGURE 10.31. Two superimposed pulses, each of a different frequency, are shown in the top photograph (A) as they enter a tube in which the sound velocity depends upon frequency. This dependence of group velocity upon frequency in such a tube causes the two pulses to become more and more separated in space (and time) as they move along the tube (B and C).

waveguide in which waves of different frequency travel at different velocities; Figure 10.31(B) shows the two pulses part of the way down the guide, starting to separate; and Figure 10.31(C) shows a continuing separation at a point further down the guide. The pulse-compression concept can be looked upon as the phenomenon shown in these figures in reverse, with the widely separated energy of part C (corresponding to the long sonar pulse of Figure 10.29) being coalesced into the superimposed pulses of part A (corresponding to the short pulse of Figure 10.29).

At about this same time, at the Murray Hill Bell Laboratories, B. Oliver (whose recent contribution to laser and hologram speckle has been referred to), issued a memorandum whose title was reminiscent of the well-known quote from T. S. Eliot's 1925 poem *The Hollow Men*: "Not with a bang but a whimper." The outgoing, extremely high-power radar pulse was generally referred to as the "bang," so Oliver's memorandum, referring to the frequency variation in the new pulse-compression concept, was entitled "Not with a bang but a chirp." But it remained for S. Darlington of the New York Bell Laboratory to fully work out the chirp details, and it is he who is given the most credit for the development of the technique.[32] Figure 10.32 shows how effective the process can be.

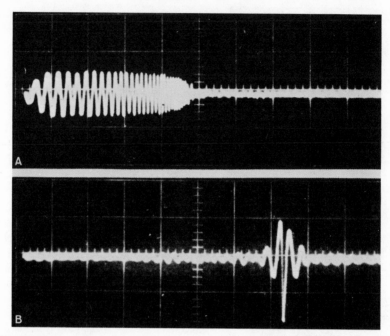

FIGURE 10.32. When the long acoustic pulse in which the frequency varies with time (top) is passed through a device in which the velocity depends upon frequency (as in Figure 10.31), the length is appreciably compressed (bottom).

In the new hologram technique, the *amplitude* of the long pulse is varied, instead of its frequency, by giving it an envelope corresponding to a one-dimensional zone plate. The length of the pulse is equal to the total length of the zone plate, i.e., the constant-frequency, continuous-wave pulse is given an amplitude-modulation pattern corresponding to that of the zone plates portrayed in Figure 10.11.

In this process, the received echoes are photographically recorded on moving film, with the echo from each target generating its own photographically recorded zone-plate pattern. As in the synthetic-aperture procedure of Figure 10.11, optical processing, with a laser, generates focused points of light (as in the right part of Figure 10.11), with each focused point corresponding to a particular target. Thus, even though the zone-plate pulse is quite long (as in the left part of Figure 10.11) the focused point of light is small, so that good range resolution is again provided, as in the chirp procedure. The range resolution for a reflecting point is equal to the smallest zone-plate dot spacing that can be resolved. The ratio of the resolution to the pulse length is referred to as the pulse-compression ratio, and with this method compression ratios of 5000:1 are easily obtained.

When there are many reflectors at various ranges, many received zone plates would be generated. These would be superimposed on the (single-line) photographic record, so that the line would comprise numerous superimposed, one-dimensional zone plates. As in a hologram or synthetic-aperture record, this superposition is nothing to be concerned about since the laser processer clearly reconstructs the individual targets.

This amplitude-modulation procedure is particularly adaptable to synthetic-aperture systems, since, for these, optical-processing techniques usually are employed anyway. To increase the energy in the outgoing pulse of such systems, the pulse can also be made quite long and be given an envelope corresponding to a zone plate. For the case of one reflecting point (as in Figure 9.21) where the usual record is a one-dimensional zone plate with very little vertical (range) dimension (Figure 9.23), the record would now appear as shown in the lower part of Figure 10.33. It would acquire an extended vertical dimension, thereby becoming a two-dimensional zone plate as shown. Optical-processing procedures would first collapse the vertical dimension, of this pattern into the one-dimensional pattern shown at the top of Figure 10.33, following which, the normal holographic processing would be applied. With many targets at different ranges and azimuthal positions, many two-dimensional zone plates would be superimposed but, as in a normal hologram,

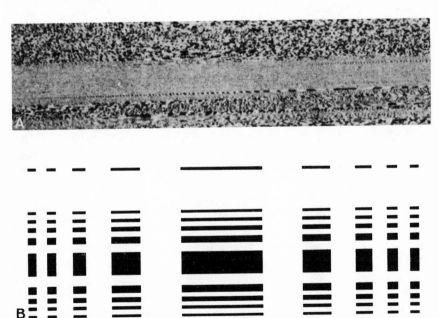

FIGURE 10.33. If a synthetic-aperture pulse (top) is given a zone-plate amplitude modulation, a reflecting point generates a two-dimensional zone plate (bottom).

optical processing would provide full retrieval of information.[33] A hologram procedure using chirp (linearly frequency-modulated) pulses has been described by Leith.[34]

10.6. Hologram Sonar Using Incoherent Illumination

In the usual procedure for making a hologram, and in hologram sonar techniques, highly coherent, single-frequency waves are used for illuminating the object and for the reference wave. In some radar and sonar applications, however, a broad-bandwidth transmitted signal is more desirable than a single-frequency one. Noiselike signals are less easily detected and are more difficult to jam or to home on. They also permit higher power to be radiated when power limitations exist in any of the transmitter components. This, is for example, the usual reason for using wide-band "chirp" signals. Recently, a procedure was described [35] in which a random noise source is used in place of the coherent waves normally used in hologram radar or sonar systems. We noted that in a continuous-wave bistatic radar or sonar system, a target P crossing the line between transmitter and receiver at uniform speed, (Figure 9.40), generates a signal (Figure 9.39) which is a one-dimensional zone plate. If, in Figure 9.40, the coherent transmitter signal T is replaced with a noise signal, the unprocessed combined signal at R would obviously not exhibit such a pattern. However, when this combined signal is given a time-frequency, narrow-band analysis, its presentation then portrays a multiplicity of contiguous, one-dimensional zone plates. The filtering can be accomplished with a large group of contiguous, narrow-band filters, or with a single, varying-frequency filter, i.e., using the visible speech procedures of R. K. Potter [36] or using fast Fourier transform (FFT) procedures.[37] (All three procedures fully utilize, of course, all of the energy in the band.) An example of the second procedure is given in the acoustic spectrogram in the upper part of Figure 10.34. For the figure a noise source was combined with a delayed replica of itself, of opposite polarity, with the delay (positive and negative) being varied at a constant rate. It is seen that the interference fringes generated by the higher frequencies of the noise signal are more closely spaced than those at lower frequencies. Figure 10.35 portrays an FFT spectrogram of a changing underwater acoustic interference pattern (from Reference 38).

Had the geometry of Figure 9.40 been used, the patterns of Figure 10.34 would have been one-dimensional (horizontal) zone plates instead of the equally spaced patterns shown, and any one of these horizontal line zone plates could then be used to reconstruct the echoing target. The signal-to-noise gain associated with the holographic (synthetic-aperture) procedure

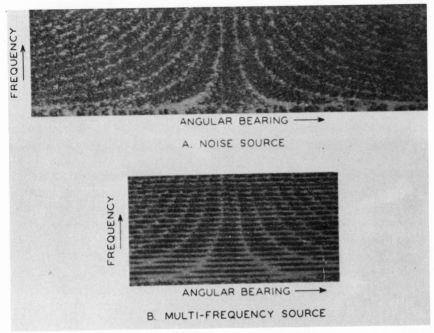

FIGURE 10.34. When a noise is made to interfere with itself and the combined signal analyzed with a narrow filter, the narrow-band filter output behaves like a *coherent* signal, exhibiting for each frequency, constructive and destructive interference "fringes." Frequency is plotted vertically, time horizontally.

would thereby be realized, whereby the entire line of the zone plate, extending over a time period during which the target has traversed a sizable distance, would be utilized (coherently). If, in addition, a conical prism lens is used, designed to focus the laser light diffracted by *all* of the horizontal line zone plates to the same focal point, a further signal-to-noise gain results.

A coherent variant of the above, again useful when power limitations exist in the transmitter components, is that of employing a multiplicity of coherent, single-frequency signals (separated in frequency). In this case the original, coherent, oscillator signals would be used to generate (e.g., by synchronous demodulation) the multiple, holographic, zone-plate interference patterns (Figure 10.34, bottom). The filtering process is thereby avoided, but the special, conical prism lens would still be required. The technique could be useful in geometrics other than the bistatic case (where coherence length requirements are quite modest), including the standard (moving) synthetic-aperture hologram systems and stationary systems operating against moving targets.

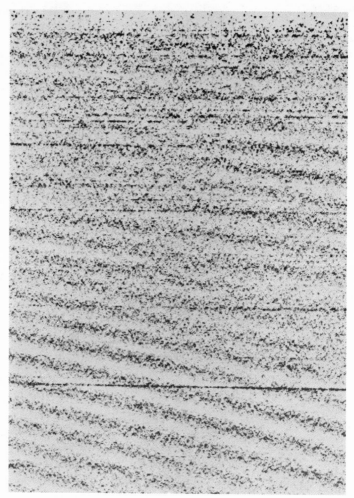

FIGURE 10.35. An FFT spectrogram portraying an underwater sound interference pattern (from Reference 38).

10.7. Seismic Holography

In sonic geophysical prospecting, low-frequency sound waves are sent into the earth, and the returning echoes are analyzed to appraise the likelihood of oil or gas being present in the substructure. Earlier, this analysis was made by inspecting the recorded, multiple-receiver traces that resulted. More recently, optical-processing techniques have been introduced in such forms as spatial filtering in order to suppress artifacts that would otherwise obscure the desired information. It is interesting to note that this particular development in seismology benefited from the optical-processing concepts that originated with the early synthetic-aperture radar work. Thus, several authors of early papers on optical seismic data processing had been contributors to the development of such radar.[39] Now the cycle has run full course, and the concept of seismic holography has been patented.[40] When this form of holography is fully developed, it should give the viewer a better view of the subterranean geologic structure and afford a higher probability of success to those in search of oil and minerals.

The usual seismic procedures use techniques which are similar to the sonar depth-recorder methods used in oceanography. In such depth recorders, a pattern is presented to the oceanographer, in a permanent record, of the variation in depth along the course that his ship followed at the time. Such a record is shown in Figure 10.36. It shows a depth profile in the Black Sea, south of Russia and north of Turkey, along one line of bearing (the course taken by the ship at that time). The ship which took this record was the Woods Hole Oceanographic Institution's research vessel, the "Atlantis II;" in the figure, the record is being examined by Woods Hole scientists. The vessel itself is shown in Figure 10.37, with the eastern shore of the Black Sea observable in the background.

In seismic applications, similar sonar principles are used to propagate sound pulses into the earth's surface by specially designed sound sources.

FIGURE 10.36. Sonar depth-finder records taken in the Black Sea.

FIGURE 10.37. The Woods Hole
vessel Atlantis II in the Black Sea.

Records of the echoes of these sounds as they are reflected from various
underground layers of rock or sediment produce profiles which often show
numerous sedimentary layers corresponding to the single bottom reflection
of Figure 10.36, and their shape can give information regarding the possibility
of oil and mineral deposits being present (Figure 10.38). Thus, a rounded or
dome-shaped layer quite often indicates an oil (and gas) deposit.

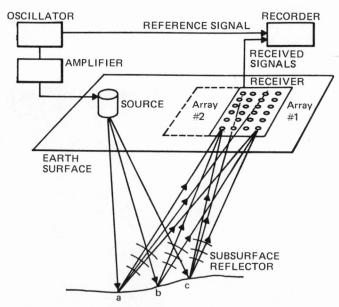

FIGURE 10.38. One possible arrangement of a seismic photographic system
(after Silverman[40]).

As has been noted, pulse-compression techniques in sonars can maintain a high resolution capability, even with long pulse lengths (high pulse energy).

In seismic applications, the chirp procedure is also useful and has been employed fairly extensively. Because high-frequency sound waves are highly attenuated in the earth, the frequencies used *are* in the very-low-frequency region (from about 10 Hz to 100Hz). However, it is difficult to build high-power transducers at these frequencies, so that many seismic sonars have embraced the chirp concept, permitting high-power pulses to be employed without impairing the resolution capabilities.

Recent field tests suggest that the technique of holography may improve the acoustic seismic process by more definitely indicating underground formations.

So far, the application of holographic techniques to large-scale underground viewing is still in the initial development stages. The frequency range employed (10–100 Hz) is so low (and wavelengths so great) that detector arrays must be extraordinarily long. However, scanning techniques, such as those used in synthetic-aperture sonar, should be useful since the objects under study are immobile.

For *offshore* seismic exploration, a system as depicted in Figure 10.39 involves a cable, which in practice would be 100 wavelengths or longer. The cable is towed behind a ship equipped with a high-power transmitter capable of emitting low-frequency, coherent acoustic energy into the ocean depths. Signals reflected or scattered from the ocean bottom, or from the geophysical layers below, would be picked up by the cable array. Holographic processing of seismic data obtained from large arrays of scanned hydrophones should provide a maximum of information retrieval from such acoustic signals.

For seismic exploration on land, the interpretation of holographic data

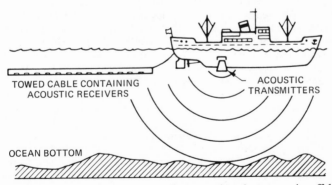

FIGURE 10.39. Acoustic holography offers certain advantages in offshore geological exploration.

from such a complex medium as the earth poses many difficulties. Inhomogeneities in the earth range from formation changes to microscopic variations in physical properties within formations (such as the change from one type of rock to another).

Nevertheless, several recent holographic field experiments have shown much promise. Thus, Lerwill,[41] using mechanical vibrators operating at 90 Hz and using a single, 4350-ft line of geophones at 50-ft spacing, was able to portray, holographically, a known, 2500-ft deep reflecting layer. Also, Fitzpatrick recently[42] conducted a holographic field experiment, imaging a known void located 80 ft below the surface of the ground.

10.8. References

1. Lord Rayleigh, *Theory of Sound*, Dover, New York, 1945, Vol. 11, p. 142.
2. Photographing sound waves, *Bell Lab. Rec.* **July**, 304–306 (1950).
3. A. F. Metherell and S. Spinak, Acoustical holography of non-existent wavefronts detected at a single point in space, *Appl. Phys. Lett.* **13** (22), (1968).
4. V. L. Neeley, Source scanning holography, *Phys. Lett. A* **28** (7), 475–476 (1968).
5. Holographic movies, *Laser Focus* **1** (17), 8–9 (1965).
6. H. Kiemle and D. Ross, *Einführung in die Technik der Holographie*, Akademische Verlagsgesellschaft, 1969, p. 17.
7. F. L. Thurstone, Ultrasound holography and visual reconstruction, *Proc. Symp. Biomed. Eng. (Milwaukee, Wisc.)* **1**, 12–15 (1966).
8. R. K. Mueller and N. K. Sheridon, Sound holograms and optical reconstruction, *Appl. Phys. Lett.* **9**, 328 (1966).
9. J. L. Kreutzer, Ultrasonic three-dimensional imaging using holographic techniques, paper presented at the Symposium on Modern Optics, New York, Mar. 22–24, 1967.
10. S. Sokolov, U.S. patent 2,164,125, June 27, 1939.
11. E. Marom, D. Fritzler, and R. K. Mueller, *Appl. Phys. Lett.* **12**, 26 (1968).
12. R. K. Mueller, E. Marom, and D. Fritzler, *Appl. Phys. Lett.* **12**, 394 (1968).
13. A. F. Metherell, Temporal reference holography, *Appl. Phys. Lett.* **13**, 10 (1968).
14. J. J. Flaherty, K. R. Erikson, and V. M. Lund, Synthetic aperture ultrasonic imaging systems, U.S. patent 3,548,642, Dec. 22, 1970 (filed Mar. 2, 1967).
15. D. F. Pekau and R. Diehl, Recording of one-dimensional holograms as a function of object range, paper presented at the International Symposium Applications of Holography, Besancon, France, July 6–11, 1970.
16. C. B. Burckhardt, P. Grandchamp, and H. Hoffmann, An experimental 2 MHz synthetic aperture system intended for medical use, *IEEE Trans. Sonics Ultrason.* **SU-21**, Jan., 1–6 (1974).
17. M. S. McGehee, Modular aperture sonar progress report, M.P.L. technical memorandum 213, Apr. 1, 1970.
18. B. P. Parkins and G. R. Fox, Measurement of the coherence and fading of long-range acoustic signals, *IEEE Trans. Audio Electroacoust.* **AV-19**, 158–165 (1971).
19. G. S. Bennett, W. E. Kock, E. J. McGlinn Jr., Internal Bendix Research Laboratories Report 1255, July 1959.

20. R. H Nichols and H. J. Young, Fluctuations in low-frequency acoustic propagation in the ocean, *J. Acoust. Soc. Am.* **43**, 716–723 (1968).

21. B. R. Brown and A. W. Lohmann, Complex spatial filtering and binary masks, *Appl. Opt.* **5**, 967 (1966).

22. A. W. Lohmann and D. Paris, Binary fraunhofer holograms generated by a computer, *Appl. Opt.* **6**, 1739 (1967).

23. B. R. Brown and A. W. Lohmann, Computer-generated binary holograms, *IBM J. Res. Dev.* **13**, 130 (1969).

24. L. B. Lesem, P. M. Hirsch, and J. A. Jordan, Jr., The kinoform: A new wavefront reconstruction device, *IBM J. Res. Dev.* **13**, 150–155 (1969).

25. J. A. Jordan, Jr., P. M. Hirsch, L. B. Lesem, and D. L. Van Rooy, *Appl. Opt.* **9**, (1970).

26. W. E. Kock, Experiments with metal plate lenses for microwaves, Internal Bell Telephone Laboratories memorandum mm 44-160-67, Mar. 27, 1944 (secret; since declassified).

27. R. J. Collier, C. B. Burckhardt, and L. H. Lin, *Optical Holography*, Academic Press, New York, 1971, pp. 345–351.

28. W. E. Kock, Metal lens antennas, *Proc. IRE* **34**, 828 (1946).

29. W. E. Kock. *Appl. Opt.* **11**, 1653–1654 (1972).

30. A. L. Boyer, P. M. Hirsch, J. A. Jordan, Jr., L. B. Lesem, and D. L. Van Rooy, Kinoform mirror for acoustic imaging, IBM Publ. 2220–6100, June 18, 1970.

31. F. Tuttle and W. E. Kock, A holographic pulse compression technique employing amplitude modulation, *Proc. IEEE* **58** (1), 170 (1970).

32. J. R. Klauder, A. C. Price, S. Darlington, and W. J. Albersheim, The theory and design of chirp radar, *Bell Syst. Tech. J.* **39**, 745 (1960).

33. W. E. Kock, Holographic amplitude pulse compression for synthetic aperture radar, *Proc. IEEE* **58**, 1773–1774 (1970).

34. E. N. Leith, Optical processing techniques for simultaneous pulse compression and beam sharpening, *IEEE Trans. Aerosp. Electron. Syst.* **AES-4**, 879–885 (1968).

35. W. E. Kock, Bistatic Microwave or acoustic holography using incoherent illumination, *Proc. IEEE (Lett.)* **61**, Oct. (1973).

36. W. E. Kock, *Seeing Sound*, Wiley-Interscience, New York, 1971.

37. W. E. Kock, *Radar, Sonar, and Holography*, Academic Press, New York, 1973. (FFT methods are described in this work).

38. G. D. Bergland, A guided tour of the fast Fourier transform, *IEEE Spectrum*, **6**, 41–52 (1969).

39. M. B. Dobrin, A. L. Ingalls, and J. A. Long, Velocity and frequency filtering of seismic data using laser light, *Geophysics* **30** (6), 1144–1178 (1965).

40. D. Silverman, Wavelet reconstruction process for sonic seismic, and radar explorations, U.S. patent 3,400,363, Sept. 3, 1968.

41. W. E. Lerwill, Holography at seismic frequencies, paper presented at the European Association Exploration Geophysical Conference, Venice, Italy, 1969.

42. R. K. Mueller, Acoustic holography, invited paper, *Proc. IEEE* **59** (9), 1319–1335 (1971).

LASERS IN INDUSTRY

Lasers and holography have found numerous uses in industry. We will review here a few of these applications.

11.1. Welding and Metalworking

The extremely high energy per unit volume which can be achieved by focusing the outputs of high-power lasers has proved useful in several areas. This high-energy-concentration capability is illustrated in Figure 11.1, showing how the invisible beam of a CO_2 laser creates a flash as it ionizes the air at the point at which the beam is focused (United Aircraft).

The use of lasers for spot welding has many advantages.[1] There is no physical contact between the welding tool and the work piece; this eliminates the need for maintenance of the usual arc welding electrodes. The welding can be done in air or vacuum because the laser beam can enter the enclosure through a transparent window. Very small welds can be made, and welds are possible in regions that are not readily accessible to conventional welding equipment. The laser beam can be controlled by a computer, by optical tracing, or by other control devices.

A particularly large and versatile metal-working laser is shown in Figure 11.2. It is used for cutting, welding, and drilling. This CO_2 laser, made by United Aircraft, puts out continuous laser power of several kilowatts and can cut, weld, and drill the hardest known materials. Similar systems are

FIGURE 11.1. An invisible beam of an atmospheric pulse laser creates a flash as it ionizes air at the point at which the beam is focused. The carbon dioxide system, which emits short bursts of light packed with extremely high energy, is capable of generating peak powers in excess of one million watts (courtesy of United Aircraft).

expected to play important roles in metal processing operations. It is also useful for cutting rock for construction and tunneling.

The Hamilton Standard division of United Aircraft, located in Windsor Locks, Conn., is known to many as one of the leaders in the production of electron-beam welding equipment. Equally notable, however, is Hamilton

FIGURE 11.2. A large CO_2 laser system used for cutting, welding, and drilling (courtesy United Aircraft).

Standard's 6-kW CO_2 laser. This unit is destined to compete with electron beams in selected welding applications. In these applications, the advantages of the laser are the absence of x rays, the use of the laser beam in the atmosphere without serious attenuation or degradation, and the simpler fixturing required through use of focusing and directional mirrors. Hamilton Standard have tended to concentrate their efforts in the field of welding, although work has been done in cutting and heat treating. They believe that the demonstrated welding performance, together with the adaptability of laser welding to automation, indicates a high potential for cost-effective processing.[2] The Hamilton Standard unit that has received considerable publicity is the fully automated, 4-kW laser welding system delivered to Ford Motor Co. This unit was designed for welding passenger-car underbodies, making three cross-body welds at a rate of 60 underbodies per hour. Its deep-penetration welds will be used on three underbody models made of rim steel up to 2 cm thick.

The system, completely automated by computerized numerical control, consists of laser beam transport and optics equipment; underbody lift, carry, and clamp mechanisms; and a fully articulated gantry that moves the laser focusing head. The laser welder delivers very little heat to the workpiece; this minimizes heat-distortion problems. The continuous-wave gas laser already has welded rim steel and high-strength alloys at rates of 500–1000 cm/min.[3]

Recently, the General Motors acquired a 1.2 kW triple-beam CO_2 laser system, used for the on-line cutting of automotive subassemblies. Producing 400 watts of power in each beam, the laser machine is the most powerful to have been scheduled for 1973 daily production use in the automotive industry. The system was manufactured by Photon Sources of Livonia, Michigan.[2]

Another, high-speed automated laser system that produces welds faster and more economically than conventional techniques has been developed, using a pulsed, neodymium–YAG, 1.06-μ wavelength laser with an average power of 150 W. It has a beam pulse rate of 100 per second and measures 58 × 56 in. high.

The laser welding system is designed for seam- and spot-welding applications. Laser manufacturers have pointed out that laser welding compares favorably with electron-beam welding in materials up to 0.5-in. thickness, being particularly suited for such applications as hermetic sealing of small metal enclosures for relays, ceramic–metal electron tubes, and solid-state devices. It can lap or spot-weld wires and microcircuit and macrocircuit leads, join dissimilar metals (such as copper to stainless steel or columbium to tungsten), and weld metals with very high melting points. The system was developed by GTE–Sylvania.[4] That company also makes a 1 kW CO_2 unit [2] designed for high cutting rates (their Model 971) and a unit (Model 1610) designed for automated seam and spot welding.[5]

Another important manufacturer is Avco-Everett in Everett, Mass. Its unit, the HPL-10, was winner of an *Industrial Research* IR100 Award in 1973. In 1974[6] the announcement was made that this 10-kW metalworking unit, for welding, cutting, heat treating, and surface alloying of metal parts, would be marketed by Sciaky Brothers, a manufacturer of electron-beam welders. Cost of complete systems were expected to range in cost up to over a million dollars, and the marker potential was foreseen as over 100 units in the U.S. and Canada. This unit, utilizing a CO_2-N_2 laser, has been extensively used at Caterpillar Tractor.[7]

Laser welding systems are also offered by Holobeam,[8] continuous (800-W YAG laser) or pulsed (50-J YAG), and the British Welding Institute has made available to institute members a 2-kW CO_2 laser welding system.[9] Hadrons Korad division offers a laser welder (Model KWD) capable of high-speed, precision, fusion welding.

A neodymium-doped glass laser was used at the U. S. Atomic Energy Commission's Fast Flux Test Facility to drill the end caps of small, stainless steel cylinders. The cylinders were then filled with a particular mixture of xenon–krypton gas and the hole sealed with a spot weld made by the same laser (Figure 11.3). These cylinders are a part of the nuclear reactor fuel pins, and if any full assembly should fail, the unique isotopic ratios of the gas mixture (the "tag"-gas) would permit rapid identification of the failed element.[10] The Fast Flux Test Facility is a major research site for the U.S. Liquid Metal Fast Breeder Reactor program; its establishment followed the successful, many-year, demonstration of the feasibility of the breeder reactor built in Idaho by the A.E.C.–University of Chicago–Argonne Universities Association, Argonne National Laboratory on the outskirts of Chicago. This program will be discussed further in Chapter 15.

A high-speed CO_2-laser drill has been in use at the Merrimack valley plant of the Western Electric Co. for drilling a complicated pattern of holes in ceramic substrates for thin-film electronic circuits. The system consists of an electronically gated CO_2 laser (with an average power output of 75 W and a peak power of 5–10 kW), a precision positioning table, and a control computer. The system was furnished by Photon Sources.[11]

11.2. Micromachining

Lasers are being used for micromachining in Germany and England. The German project uses a 3-in., water-cooled ruby laser and the applications include microelectronic bonding and thick-and thin-film resistance control. The unit has been production tested for a wide range of drilling, welding,

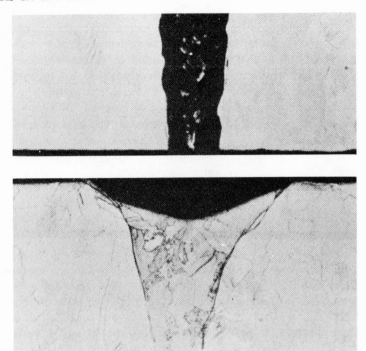

FIGURE 11.3. A typical hole produced by laser drilling (top), and a laser-produced spot weld (bottom) are shown in these views. The cylinder end cap is 20 mils thick.

cutting, trimming, and surface-treatment uses, involving metals and non-metals. Use under production conditions has indicated an operating cost of under 0.1 cent per flash, attributable in part to the use of an ellipsoidal pumping mirror, which permits extremely high utilization of the output of an inexpensive flash tube. The ruby laser combines a variable pulse length for different machining operations with constant pulse energy and a 25 per second repetition rate. Available power density is 10^9 W/cm². Short-time pulse-energy variations are automatically minimized by a voltage regulator, and long-time variations by a laser-energy stabilizer.

Microwelding possibilities include: microelectronic contact bonding, constant bonding of printed circuits, spot welding plastics and ceramics, nondeformative welding of cured microparts, welding of thermocouples and clock components, and soldering of insulated wires. Drilling possibilities, with hole diameters down to a few microns, include: diamonds as die stones, ruby disks as jewel bearings, metal disks as flow resistors, and hard metal rings as air-pressure bearings. Surface treatment possibilities include: cutting of vapor-deposition masks, thick- and thin-film resistance control, and surface engraving. In England, a He–Ne laser has been used for cutting photolitho-graphic masks, in-contact masking, or for direct deposition of interconnection patterns. Radiation from the laser is focused by a lens onto the metallic coating of a glass substrate. The intensity is sufficient to boil the metal at the point of incidence and remove it by evaporation. In a powerful British system the laser trimmer consists of a continuously pumped solid-state laser Q-switched at high frequency to produce a series of high-peak-power pulses of infrared radiation. When focused to a typical spot diameter of 0.001 in. the pulses vaporize most metal and metal oxide films on glass, ceramic, and other substances. A flow of cooling water is required.

11.3. Solid-State Trimming

A U.S. laser system for trimming thick-and thin-film resistors combines dual-probe operation with a numerically controlled xy table, and thereby eliminates restrictions on the number of resistors per pattern, the density of pattern, and the resistor orientation within the pattern. It consists of four major subsystems: the NC positioner, the control-tape programer, the probing and trimming interface, and the laser. The two probes of the trim-mer are independently programable. Probe position is stored on the control tape. On command from this tape, the xy table will automatically rotate the substrate.

The positioner is a tape-controlled, ac servo-driven, dual-axis table with full-scale travel in each axis of 10 in. Table-positioning speeds of up to 25 in./sec are provided and available. The positioner provides three modes of operation: manual, semiautomatic, and automatic.

A laser solid-state machining system described by the Bell Telephone Laboratories is capable of forming tiny electronic circuit patterns directly onto ceramic substrates, or circuit bases, in one simple step. Earlier methods for producing precise circuit patterns on ceramic required clean-room con-ditions and many processing steps, including mask-making, photoresist application, and chemical etching. The new process makes use of a laser assisted by information stored in a computer that is programed to describe the type of circuit pattern to be machined. An important application for ceramic circuits formed by the laser machining process is the interconnection of electronic devices such as silicon integrated circuits. Interconnection via ceramic circuits has advantages over other techniques, such as printed wiring boards, in that it permits finer-line conductor patterns, higher packing density of electronic components, and superior circuit performance. Substrates coated with a thin film of conductive metal are mounted to the outside surface of a circular drum. As the drum rotates, each substrate is successively exposed to a focused laser beam that is modulated, or switched on and off. The modulation of the laser beam allows microscopic regions on the metal coating of each substrate to be either selectively vaporized or left intact. Unwanted material is then removed along a line that extends across each substrate. After one complete revolution of the drum, a reflecting mirror "steps" the laser beam to the next line to be machined. This continues until every point needed to form circuit patterns on the substrates has been scanned by the laser. All of the metal film portions that are not vaporized constitute a circuit pattern after the machining process has been completed (Figure 11.4). Modulation of the laser beam in the machining process is achieved by a train of coded signals stored in a computer. These signals represent the electronic circuit pattern to be machined on the substrates.

In this process, most of the fabrication steps of conventional methods have been eliminated. It leaves the surface uncontaminated by organic residues. It is a noncontact process which avoids damaging the metal film. It is

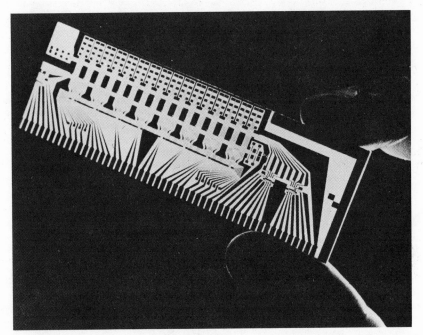

FIGURE 11.4. This exotic circuit pattern was machined by laser light, directly onto an insulating substrate. Circuits such as this one constitute the building blocks that are required in the sophisticated electron components used in Bell System switching and transmission devices, as well as in other telephone equipment and services (courtesy of Bell Telephone Laboratories).

insensitive to dust, eliminating the need for clean-room conditions and it is repeatable since there are no photographic masks to wear out.

Resistor trimming systems are now available from many U.S. companies. Hadron-Korad of Santa Monica, Calif. offers a modular system which offers three capabilities, a manual one for laboratory work, a semiautomatic one, and a fully automatic one.[13] It permits tight tolerances to be held in production trimming of both thick- and thin-film resistors, with trimming accuracies of several parts per million. Figure 11.5 shows, the various-thickness kerfs possible with this trimmer. Production rates of several thousand resistors per hour are possible. The depth of the focal area of the laser is purposely made ± 5 mils; this results in an essentially constant kerf even if the resistor films should vary in thickness or the substrate vary in flatness.

The advantages of laser trimming include sharply defined sealed edges, and the satisfactory vaporization of gold, cermet, or other types of resistor compositions.

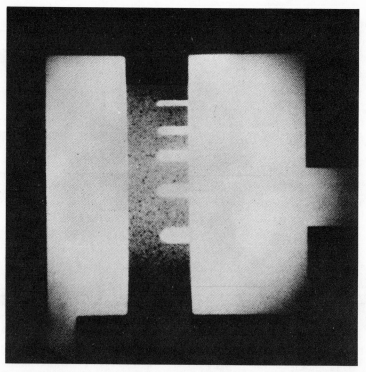

FIGURE 11.5. Various kerfs on the thin film (Ni–Cr) are cut at constant power density (35× magnification). The smallest kerf is 0.8 mil (courtesy of Hadron Corp.).

Other resistor trimmers are made by Arvin,[14] Electro-Scientific Industries,[15] and Teradyne.[16]

11.4. Laser Scribers for Substrates

In today's solid-state circuitry, the components, as we saw in Figure 7.13 are extremely tiny. The circuits are made by photographically reducing artist's drawings to a tremendous degree, and causing the circuits to be deposited on a flat support called a substrate. Since the photographic replication of the artist's drawing of a given circuit (such as the one shown in the eye of the needle of Figure 7.13) is a simple process, the usual procedure is to place perhaps hundreds of devices (circuits) on the same wafer substrate, with the substrate usually being ceramic, glass, or sapphire. The problem then, is to divide the rigid, brittle wafer substrate into the individual circuits (devices). For a long time the wafers were scratched with steel or diamond scribers and then broken into the individual elemental devices; this procedure often resulted in many broken devices.

The focused laser provides an excellent way for scribing the wafers to exactly the depth necessary to eliminate broken units, and today there are a number of such scribers available on the market. One model, offered by Electro-Scientific Industries, uses a 50-W CO_2 laser operated in the pulsed mode, with the process controlled by a DEC PDP-8 minicomputer.[17] A model offered by TRW uses a TEA laser.[18] A model offered by Hadron-Korad follows the usual pattern of having the operator position the wafer by viewing it through a microscope; it embodies, however, a split-field microscope, providing the operator with two large fields of view and two cross hairs to permit accurate angular, vertical, and horizontal alignment (Figure 11.6). Figure 11.6 also shows the many individual solid-state circuits. The unit also has an automatic staging table that is programable and therefore does not require computer or punched-tape input.

The Japanese government's Electro-technical Laboratory has described an experimental laser wafer scriber which uses the laser for positioning also. The laser is used to sense the relative position of the wafer, and when it locates the pattern edge, the input information enables the system to find the position of the starting corner.[19]

11.5. Microdrilling

Lasers have also found application in several industrial microdrilling operations. Thus, in the past, adjusting the frequency of a clock's balance

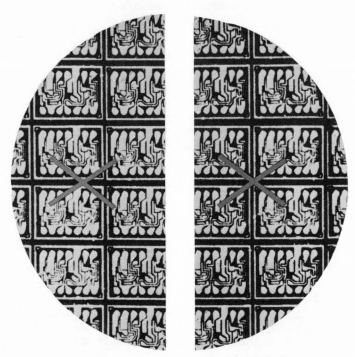

FIGURE 11.6. Final alignment of a wafer, rotated to proper angle θ and proper coordinates (x, y), for start of scribing (courtesy of Hadron Inc.).

wheel was a tedious process of handdrilling holes to reduce the wheel's mass, and then checking for balance.

Using a dual-beam pulsed YAG laser, the job can be performed neatly in a tenth of the time it used to take. The laser drills the holes automatically while the wheel is rotated on a precision shaft (Figure 11.7). The laser beam exerts no appreciable force on the wheel, so there is no distortion.[20] In England a 70-J ruby laser has been used for "microdrilling" extremely small holes in nuclear reactor components so as to simulate small leaks. Lasers are also being used to manufacture pinholes in stainless steel disks for use in optical devices such as spatial filters. Figure 11.8 shows a laser drilling a hole 0.001 in. in diameter in the pointed end of a serving needle, at the Arnold Engineering Development Center of the U.S. Air Force Systems Command.[21] The needles are then used in a study to improve the construction of a hot-wire anemometer for use in wind tunnels to measure airflow velocities. The drilling capabilities of lasers are also used to remedy flaws in diamonds. A solid-state YAG unit is used to drill tiny holes down to the imperfections,

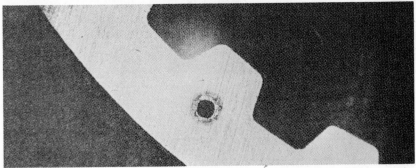

FIGURE 11.7. Neodymium–YAG laser puncturing two holes in a clock balance wheel (top). A close-up is shown at the bottom (courtesy of Bulova Watch–Sylvania).

which are then removed by a variety of methods.[22] The need arises because natural diamonds often contain such imperfections as "voids" (gaps in the crystalline structure), "twins" (variations in crystal growth from the natural, single-crystal structure of the stone), and "inclusions" (the presence of foreign material, both crystalline and noncrystalline, within the stones). After a stone is faceted, these imperfections, particularly inclusions, may show up darker than the surrounding material, thus detracting from the stone's natural brilliance. Since these imperfections are within the stones, in the past there was little that could be done to mask or remove them. The holes pierced by the laser are 0.002 in. in diameter, about the thickness of a human hair. They are not generally visible without magnification, and their diameter is so small that they do not change the stone's natural reflections. Laser-drilled holes in diamond have been used for several years for wire-drawing dies. Such

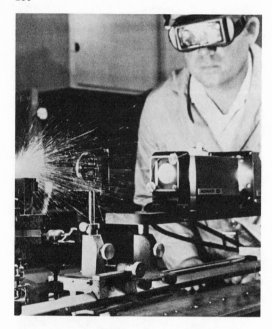

FIGURE 11.8. A 0.001-in.-diameter hole being drilled in the pointed end of a serving needle by a laser (courtesy of U.S. Air Force Systems Command).

holes were previously made by steel drills, a process taking four or five days.

Finally, the microdrilling of extremely small holes in contact lenses appears promising. These holes allow air to pass through to the surface of the eye, which stimulates the circulation of natural fluids around the eye and retards the onset of itching (which many wearers experience). But because of the expense and difficulty of drilling the holes by hand and then cleaning them, this process has only been used as a last resort. In the laser process, a 50-W CO_2 laser is used. The process can be automated and there is no need for subsequently cleaning the holes. Experimentation has shown that 60 holes, each 0.1 mm in diameter, provide the maximum comfort.

A September 1974, advertisement described Western Electric's use of lasers in the microdrilling of ceramic substrates. In October 1974, Parker pens advertised "Laser-Engraved Wood-Cube Pen Sets."

11.6. Fabric Cutting

Interest in the use of lasers for cutting fabrics has grown rapidly in the last few years, with numerous apparel manufacturers purchasing cutters, each costing over a half million dollars. The advantages quoted are the ability to quickly respond to fashion changes, the better fitting clothes which result, and the cutting to very close tolerances without error, resulting in less waste.[21] One such machine is shown in Figure 11.9. In 1973 it was announced that Richman Brothers of Cleveland, Ohio (a division of F. W. Woolworth and a manufacturing retailer of men's tailored clothing) had purchased two cutters and that C. Itoh, a Japanese trading company, had also purchased two cutters for delivery to a Japanese apparel manufacturer, this being the first sale outside the United States.[23]

11.7. Checkout Systems for Supermarkets

A development that is expected to be at the checkout counters "in most markets by 1980"[24] involves the laser "reading" of a code stamped on packages, and the transmission of the information to computers for rapid determination of the total bill. The computers tabulate the data into inventory information, prices and sales taxes, and deliver to the customer the totals, his change, and any trading stamps. The checkout attendant, freed of cash-register duties, can immediately go about placing the merchandise in bags. Savings in productivity will total 1–1.5% of annual sales, according to

FIGURE 11.9. A fabric-cutting system, using a laser for cutting (courtesy of Hughes Aircraft Co.).

representatives of the seven major U.S. grocery associations.[24] Because supermarket profits are presently only about 1% of sales, such a saving would be quite significant.

One of the first experiments in testing this development was conducted at a Kroger supermarket at the Kenwood Plaza Shopping Center in Cincinnati, Ohio, starting in the summer of 1972. This store has annual sales of 3.7 million dollars. The checkout system used was developed by the RCA Computer Systems Division (Marlboro, Mass.) and uses the circular code shown in Figure 11.10. A low-power laser beam is expanded in an optical system and caused to strike a rotating multifaceted mirror. The mirror scans the beam across a slot in the checkout counter and thus scans the code on packages slid over the slot. The reflector beam conveys the information to a detector, which then feeds it to the computer for processing. At the Kroger store in Kenwood Plaza, customers turn their packages so the code is on the bottom. A checker scans and bags items in one continuous motion. As each package passes over the slot, a "beep" confirms that the price has been recorded and is displayed at a window facing the customer (Figure 11.11). If a beep is not heard, the checkout clerk returns the package to the slot. The test has been a striking success according to Kroger's director in industrial engineering, Robert L. Cottrell, who noted that more than three million items were scanned without a mistake.[24]

FIGURE 11.10. A code affixed to a grocery store item can define the product.

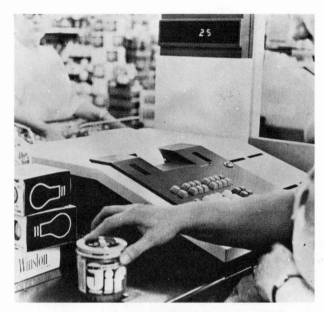

FIGURE 11.11. At a Kroger Store in Cincinnati, a clerk passes jar across scanner while customer empties his cart. The price is computed and displayed automatically (top).

More recently, IBM [25,26] and National Semiconductor [27] have introduced similar checkout systems. Other supermarket scanners are built by the Singer Co., National Cash Register Co., and Litton Industries, Inc. [24,26] The grocery industry has now adopted a standard (universal product) code of identifying symbols for these checkout-counter laser scanners. The Bendix Corporation also developed a laser label reader which can scan a label, and thus identify objects, on a conveyor line. [28]

11.8. Display Applications

We discussed in Chapter 5 how the amazing three-dimensional realism of hologram images resulted in wide publicity for a very dramatic hologram (Figure 5.2) and for the store (Cartier's) which displayed it. Because visual appeal is the basis for most advertising and commercial displays, the potential of lasers and holographic techniques for creating new visual forms and for enhancing our perception of old forms has not gone unnoticed. And because holographic images are three-dimensional and can be projected so as to appear free-standing, they have more dramatic possibilities than any two-dimensional medium.[29]

Engineers at the Fisher Body Division of General Motors incorporated several developments in holography into a product display at the General Motors Building in New York City. Visitors to the exhibit peer through plate-glass "windows" mounted in four sides of an octagonal display case. Each "window" is actually a hologram containing a record of two different scenes. In ordinary light, the holograms appear to be transparent. When a special high-intensity lamp illuminates the hologram surface, a three-dimensional image of the Napoleonic coach (Fisher Body's symbol) becomes clearly visible in the center of the exhibit case (Figure 11.12). After 15 seconds, the coach disappears and the detailed image of an automobile body appears in its place (Figure 11.13). The images repeat this cycle, with only a slight break to light the vacant interior and prove that no real models are used for

FIGURE 11.12. Two photographs (in the two corners) of the reconstructed hologram of the Fisher Body Napoleonic coach. In the middle is a series of photographs also showing the three-dimensional nature of the image, as displayed in the General Motors Building in New York City.

FIGURE 11.13. Arrangement for recording the General Motors hologram of Figure 11.12. To record a second image on each plate, the coach was replaced by an automobile body, and an alternate reference beam was used to expose the photographic emulsion.

deception. To lend another element of realism to the display, individual holograms were recorded of side views as well as the front and rear perspective of both the coach and auto body. They are positioned in such a manner that the observer can circle the display, moving from one hologram to another, and receive a 360° view of both objects. The exhibit was the first commercial application of integrated holograms in a product display, and its single-source lighting system illuminating all four holograms simultaneously was also novel.[30]

More recently, the New Museum of Witchcraft and Magic, located on Fisherman's Wharf in San Francisco, Calif., featured one of the largest holograms (65 × 95 cm). In it, a "beautiful witch" holds what appears to be a large ball, 12 in. forward of the hologram plate, in which the viewer sees another hologram reconstruction, this one of "Hades, where the devil beckons to the viewer".[31] The double, three-dimensional parallax effects caused by the *two* holograms are quite spectacular.

At the Fall 1973 Trade Show for Trade Shows Convention, held in Detroit, Mich., McDonnell-Douglas exhibited an impressive group of large holograms, including a 3-ft-diameter cylinder that appeared to contain a DC-10 jet. Actually, the cylinder was empty, the image having been created by a concentrated light shining on a transparent band of film mounted on the cylinder. Another 2-ft × 3-ft hologram showed a "real-life" underwater scene behind it, including five scuba divers. Also included were copies of the Cartier hologram (Figure 5.2) and the witch and devil hologram just described.[32]

One form of hologram display has been exploited by TRW under the direction of Dr. Ralph Wuerker. Holograms have been made of numerous, famous, centuries-old statues in Venice so that if eroding or settling continues, a three-dimensional record of their outlines at the present time will have been recorded.[33] TRW has also collaborated with the U.S. Air Force Systems Command to display the spray pattern of a liquid propellant in a rocket engine (Figure 11.14). The distribution of propellant droplets is a very important factor governing the performance of the engine, and it is also the most difficult variable to measure and predict.[34]

Finally, in the display category, the use of lasers of three colors permits large-scale television pictures to be projected. Japan's Hitachi Ltd. uses a krypton ion laser to generate the red primary color, an argon ion laser for

FIGURE 11.14. Spray pattern of liquid propellant is analyzed at U.S. Air Force Laboratory.

the green, and an argon laser for the blue. Scanning is accomplished mechanically, using a 25-facet rotating mirror for the horizontal and a vibrating galvanometer mirror for the vertical.[35] RCA has demonstrated the use of an acousto-optic deflector to scan a laser beam at television rates.[36] Still another approach was described by the 3M Company.[37] There the lasing surface was a flat plane; when a point of the plane was struck by an electron beam, laser light was emitted from the opposite side (as in the case of the phosphor screen of the usual cathode-ray tube).

11.9. Cable Insulation Stripping

Several companies have introduced cable-stripping industrial laser systems. One is a complete material-processing system that utilizes an industrial CO_2 laser to remove simultaneously insulation from both sides of flat bus cable. Others can slit the insulation along the wire, permitting it to be easily removed. Such stripping by lasers offers the following features: No nicks, scratches, or cuts are produced in the metal conductor or sheath where the separation of the insulation is performed; the metal conductor is uncontaminated; and bundles of braided wires can be laser-stripped without uncoiling each wire. Lasers have been used in this application by cable, aerospace-wiring and computer manufacturers.[2]

11.10. Control of Crystal Growth

In Germany, the growth of synthetic crystals has been controlled more precisely than is possible with conventional seed-pulling and crystal-rotating schemes. With the new method, which is based on laser techniques, the diameter of the crystal throughout its growth can be measured and controlled with an accuracy to within 1%. It thus solves the problem of irregular diameter variations, which in normally grown crystal can give rise to structural imperfections. The laser method exploits the surface tension effects that cause the free liquid surface of the melt from which the crystal is grown to curve upward near the crystal. The surface inclination at a fixed point close to the crystal is a measure of the crystal diameter. The magnitude of this slope is then detected by the laser beam and measured with an arrangement consisting of an optical reflection system and a pair of photodiodes.

11.11. Chromatograph-Analysis Use

Gas-chromatographic analysis of plastics and other organics has been enhanced and speeded up with a laser technique that cleanly and almost instantly fragments the test samples. With the new method of preparing materials for analysis, both quantitative and qualitative analyses have been made in a fraction of the time required using older techniques such as pyrolysis or high-voltage electric discharge. When put into the fragmentation cell, the material can be positioned in a nearly evacuated space where the laser, a neodymium-doped glass type, almost instantaneously decomposes it. Small, easily decomposed polymer and copolymer fragments are obtained, whereas tars and by-products are reduced to a low level, simplifying the analysis. Components vaporized by the laser are transferred immediately by an inert carrier gas to the gas chromatograph column for separation and identification, permitting the transfer to take place quickly so as to prevent the vaporized components from settling out. Chromatograms made using laser-fragmented material were characteristic of the polymer type and were less complex than those obtained by thermal pyrolysis.

11.12. Weather Analysis

Lasers also have permitted ocean-going vessels to make atmospheric measurements for various purposes. Figure 11.15 shows a laser (a Korad

FIGURE 11.15. A laser on an oceanographic ship used for atmospheric studies (courtesy of Hadron Inc.).

K-1Q laser) mounted on the fantail of an oceanographic ship used for atmospheric studies.

11.13. References

1. R. D. Haun, Jr. Laser applications, *IEEE Spectrum* **5**, May, 82–92 (1968).
2. G. E. Overstreet Heavy-Duty light, *Ind. Res.* **16** (5), 40–44, (1974).
3. *Ind. Res.* **14** (11), (1972).
4. *Res. Dev.* **24** (9), (1973).
5. *Opt. Spectra* **7**, Sept., 54 (1973).
6. *Electron. News* **18**, June 10, 49 (1974).
7. *Opt. Spectra* **7**, May, 40 (1973).
8. *Laser Focus* **10**, Feb., 41 (1974).
9. *Lasersphere* **3**, Jan. 15, 3 (1973).
10. *Opt. Spectra* **7**, Apr., 19–20 (1973).
11. *Opt. Spectra* **8**, Apr., 46 (1974).
12. *Microwaves* **10**, Mar. (1971).
13. *Lasersphere* **3**, May 15, 26 (1973).
14. *Lasersphere* **2**, Dec. 15, 16 (1972).
15. *Electron. News* **18**, July 23, 67 (1973).
16. *Electron. News* **19**, Apr. 15, 74 (1974).
17. *Electron. News* **19**, Aug. 19, 74 (1974).
18. *Laser Focus* **7**, Mar., 12 (1971).
19. *Electronics* **46**, Aug. 30, 52 (1973).
20. *Microwaves* **9**, Sept. (1970).
21. *Gov. Exec.* **July** (1971).
22. *Opt. Spectra* **6**, May (1972).
23. *Opt. Spectra* **7**, Sept. 12 (1973).
24. Revolution at checkout counters is built around a laser scanner, *Laser Focus* **9**, Aug. 10–12 (1973).
25. Laser checkout system rings groceries up faster, *Lasersphere* **3**, Oct./Nov., 1–3 (1973).
26. IBM builds scanner for grocery stores, *Laser Focus* **9**, Nov., 4 (1973).
27. *Electron. News* **19**, Apr. 15, 66 (1974).
28. *Ind. Res.* **14** (1), 14 (1972).
29. A. L. Hammond, Holography: Beginning of a new art form, *Science* **180** 484–485 (1973).
30. Holographic display featured in G. M.'s new international building, *Laser Focus* **4**, Oct., 18 (1968).
31. *Holosphere* **2**, July, 8 (1973).
32. *Holosphere* **2**, Oct., 15 (1973).
33. *Microwaves* **11**, May, 22 (1972).
34. *Laser Focus* **6**, June, 20 (1970).
35. *Lasersphere* **3**, June/July, 6 (1973).
36. *Laser Focus* **9**, Nov., 30 (1973).
37. *Sci. News* **May 4**, 286 (1974).

COMMUNICATIONS

Lasers have been widely used in the field of communications. In this chapter we will discuss some of these applications.

12.1. The Constant Demand for Bandwidth

The history of communications technology is replete with examples of successful efforts aimed at increasing the useful bandwidth (the width of the frequency band required to transmit a given type of signal) of communications circuits. The proven value of telegraph messages (dots and dashes at quite slow repetition rates with a bandwidth requirement of only a few hundred hertz) led to the development of the telephone, which even today requires, for effective voice transmission, a bandwidth of only 4000 Hz. For high-fidelity systems, enthusiasts insist on an audio bandwidth equaling our full hearing range, extending from about 30 Hz to 15 or 20 kHz. For many years, telephone communication circuits comprised two wires, which later became bundles of twisted pairs, these providing more phone circuits (more bandwidth) for a given length (separation distance).

12.2. Coaxial Cable

An extremely significant advance occurred with the development of the much broader bandwidth *coaxial cable*, invented by Lloyd Espenscheid of the Bell Telephone Laboratories. This cable, along with the *carrier* system, whereby many phone circuits are stacked on top of each other, eventually enabled hundreds of thousands of phone conversations to be transmitted over a single circuit. At the 1939 World's Fair at New York City, U.S. television was shown widely for the first time. It was obvious that *network* television, comparable to the network radio of many years use, would become a highly desirable and very important possibility. The coaxial cable *did* provide such network TV for a while, but the cables of those days, with their 2.5-MHz capabilities, could not transmit the full 4.5-MHz TV signals. The black and white pictures were fuzzy and color TV was completely out of the question.

12.3. Microwave Systems

The next advance came with the microwave waveguide, developed approximately simultaneously by a group at the Bell Telephone Laboratories under C. G. Southworth and by another at the Massachusetts Institute of Technology headed by W. L. Barrow.[1,2] The waveguide concept was difficult for many to understand, since a waveguide is a single tubular conductor. The announcement of the transporting of electrical energy over a single conductor, a hollow metal tube, caused many questions to be asked. Where is the return circuit? If the current travels out in the tubular conductor, how does it get back? Of course when waveguides were described to physicists versed in optical matters, no such consternation was evident. They recognized that radio waves and light waves, both electromagnetic, would behave alike, and the idea of passing light waves down a tube with mirror-like walls seemed not at all radical. The radio waves must simply have wavelengths sufficiently short so as to behave like light waves. Waveguides played an important part in the development of microwave radar[3] during the early 1940's. Thus, waveguides were involved in Figures 2.5–2.7, 3.7, 9.9, 10.23, 10.24, 10.27, and 10.28.

12.4. Radio Relay Circuits

An equally important application occurred in the communications field in the development of microwave radio relay circuits.[4] The relay stations (Figure 12.1 and 12.2) used the waveguide lens of Figure 5.7 and the strip lens of Figure 10.3 to span the United States with bandwidths far exceeding the then existing coaxial cables, permitting network television to provide color TV pictures to much of the nation. Japan followed shortly thereafter[5] with a microwave relay circuit (Figure 12.3). These circuits utilized tall towers (such as the one shown in Figure 12.2) because of the need for estab-

FIGURE 12.1. A tower in the first link, the New York-to-Boston link, of the Bell System's first transcontinental microwave relay circuit. Of the four horn-lens antennas, one receives a signal and a second transmits it onward, for one transmission direction; the other two handle signals transmitted in the opposite direction.

lishing a "line of sight" for the very-high-frequency radio waves, comparable to the need for a clear line of sight for light waves. The earth's curvature thus places a limit on the separation of two tower-borne microwave antennas, with the possible separation being larger the higher the towers. Accordingly, although the Bell System's first transcontinental microwave radio relay developments (Figure 12.4) were extremely important for overland routes, they could not be used in spanning the oceans. For such applications, only the older, undersea, coaxial cables were usable.

12.5. Satellite Communications

It was, therefore, not surprising that when the U.S. space program began showing its mettle, with its more and more powerful rockets, Bell System engineers and scientists began considering ways for increasing the bandwidth capacity of transoceanic circuits through the use of *satellite* microwave communication circuits. Some of this interest stemmed from a paper

"Extra-Terrestrial Relays" authored by the British engineer Arthur C. Clarke and published in 1945.[6] Clarke, in his paper, did note that rockets to achieve orbital or escape velocities were still some years away. The U.S. rocket pioneer Robert H. Goddard was moving in that direction, as were German experts under Wernher von Braun at the research center near the city of Peenemunde in northeast Germany. Toward the end of World War II, the ballistic, rocket-powered missile, the V-2, caused visionary rocketeers, both in the U.S. and the Soviet Union, to begin thinking of rockets, not just capable of sending missiles across the English Channel (the V-2), but capable of sending them across the oceans, in short, rockets for powering intercontinental ballistic missiles (ICBM's). The Soviet Union surprised many with their announcement of a successful test of such a missile in August 1957, and on October 4, 1957, the world was startled by the Soviet announcement of its successful launching and orbiting of Sputnik I, the world's first man-made

FIGURE 12.2. A tower in the New York-to-San Francisco route. In the flatter midwest terrain, height, to shorten the tower separation, is not available from mountains; the towers are therefore quite tall.

FIGURE 12.3. Horn-lens radio relay antennas in Japan (courtesy of Mitsubishi Electric).

moon (man-made satellite). It convinced the world of its presence by its periodic radio transmissions of audible "beeps." This Soviet feat was possible because of the existence of the more powerful Soviet rockets, developed for their ICBM program. The U.S. rockets, although adequate for U.S. ICBM's were inadequate for the task of orbiting a satellite. Development programs for more powerful rockets had been under way, but in order to gain time and speed up this development, rocket expert Wernher von Braun was called upon by President Eisenhower to embark on a crash program. It is quite likely that U.S. public opinion was responsible for this action, as was also the nation's desire to show the world that U.S. technology was fully capable of meeting the challenge. Responding to this public opinion trend, the then Governor of the State of Michigan, G. Mennen Williams, composed the following poem referring to President Eisenhower's great interest in golf:

> Oh, little sputnik, flying high
> With made-in-Moscow beep,
> You tell the world it's Commie sky,
> And Uncle Sam's asleep.
>
> You say on fairway and on rough
> The Kremlin knows it all.
> We hope our golfer knows enough
> To get us on the ball.
>   (courtesy Associated Press)

Von Braun's team *did* meet the challenge, and within four months, a U.S. "moon," Explorer I, was orbiting the earth. But the bigger, earlier, Soviet rockets continued to steal the show, and on April 12, 1961, a space ship, with Cosmonaut Yuri Gagarin on board, was orbiting the earth, even *before* the exciting, but less impressive, suborbital space flight, on May 5, 1961, of U.S. astronaut Alan Shepard, involving a bulletlike (ballistic) trajectory from Cape Canaveral to the South Atlantic.

But the U.S. space program soon reached full momentum with newer, bigger, and more powerful rockets appearing almost monthly. This great speed of development was a requisite, if President Kennedy's announced goal of landing a man on the moon by 1970 was to be achieved. As we know, that goal was achieved when U.S. astronaut Neil Armstrong became the first man to set foot on the moon on July 20, 1969.

12.6. Orbiting Satellites

For quite some time *before* that historic event took place, much thought and research had been directed toward satellite communications.[7] Figure 12.5, showing how the television relay capabilities of a satellite were envisioned, was published in 1961, and a very important point in the U.S. space development program occurred on July 10, 1962, when Clarke's concept[6] was

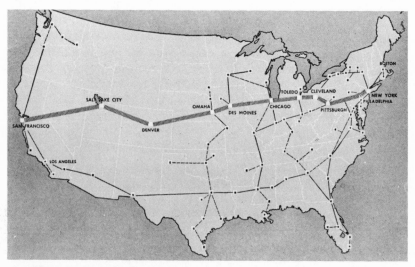

FIGURE 12.4. The dotted line is the path of the Bell System's first transcontinental microwave radio relay circuit.

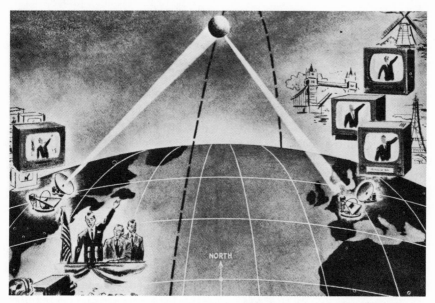

FIGURE 12.5. This sketch, published in 1961, predicted trans-Atlantic television transmission via satellite.

brilliantly demonstrated by the American Telephone and Telegraph Company, with their TELSTAR satellite, relaying color television programs between Europe and the United States. In 1967, the Soviet Union decided upon that technique for implementing broad-band communication to about two dozen cities which are quite distant from Moscow, including several in Siberia, with a system called Orbita Molina,[8] a system still in use in the Soviet Union in 1974.

12.7. Synchronous Satellites

The communications satellite which was to demonstrate the greatest promise for world-wide satellite communications, was SYNCOM. Sputnik, TELSTAR, and, in fact, most of the earlier earth-orbiting satellites, orbited the earth once every two hours or so, and the altitudes of their orbits were about 100 miles above the earth's surface. As these communications satellites thus passed overhead, the ground stations, in order to transmit signals to them and to receive signals from them, had to be equipped so that the aiming angle of their directional beams could be continuously varied (so as to always be pointing at the orbiting satellite). The SYNCOM scientists proposed elevating their satellites, not to 100 miles, but to *22,300 miles* above the earth. The key to this was that at that tremendous altitude, the orbiting time is not 2 hours but 24 hours. Accordingly, since the earth itself rotates on its axis once a day, i.e., *also* once every 24 hours, the 24-hour orbit of such a satellite causes it to appear to an observer on earth to be stationary. Because of this, it is referred to as a geostationary satellite (geographically stationary). The obvious advantage of this arrangement is that the ground stations can remain in continuous radio contact with the satellite without having to alter the pointing direction of their antenna beam patterns. If the satellite is "parked" over the Atlantic, the U.S. ground stations and the European ground stations can be *permanantly* aimed at the satellite, and the cumbersome variable pointing requirement of the ground station for the more rapidly orbiting satellites is thus eliminated.

One obvious disadvantage of the geostationary satellite is that the rocket required for elevating it to 22,300 miles must be much more powerful than the rocket for the 100-mile-altitude satellite. But the space program has furnished such powerful rockets. The first synchronous (geostationary) communications satellite, SYNCOM I, was rather light in weight and hence rather small. But its light weight enabled a then available rocket to hoist it to the needed 22,300-mile height. The success of SYNCOM I pointed toward a bright future for communications satellites, and soon the newly formed U.S. satellite corporation, "ComSat" (for Communications Satellite Corporation), was employing much larger rockets to put much larger communications satellites into geostationary orbits. By 1970, such satellites were encircling the globe (Figure 12.6), providing world-wide communications circuits (phone and television circuits)[9]

12.8. Domestic Systems

Initial application for a U.S. *domestic* satellite system was made in 1965, but in the view of Waynick, "the U.S. Federal Communications Commission is apparently following its usual practice of protecting vested interest," [8] and nothing resulted from that application for almost 10 years. Canada, however, did not need to bow to the U.S. FCC delays, and in 1972, Canada's first domestic satellite Anik I (Anik is Eskimo for brother) was launched into a geostationary orbit, and a second one, Anik II, was launched in 1973. In August 1974, RCA announced the use of Anik II to provide a U.S. domestic circuit, thereby initiating the nation's first domestic commerical satellite service connecting the East and West coasts. RCA used ground stations near

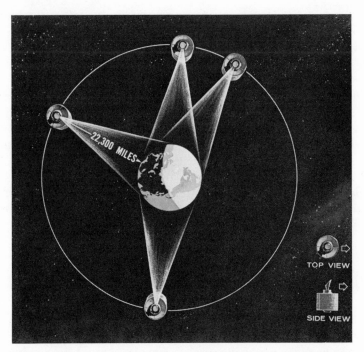

FIGURE 12.6. Synchronous orbiting satellites (stationary with respect to the earth) are now "parked" over the Atlantic, Pacific, and Indian Oceans, providing world-wide television coverage. The Intelsat III model is sketched here at the right.

New York and San Francisco. The *Wall Street Journal* article describing this development noted that "because a satellite connection can be completed with fewer relay points than can an earth-bound system, the quality of transmission is better." By 1974 the parade toward the domestic satellite communications business had included Western Union, American Satellite Corp., Communications Satellite Corp., RCA, and IBM.[10] In September 1974, the nation's first truly domestic communications satellite, WESTAR I, owned by Western Union, became operational.

12.9. Early Laser Research

Meanwhile, the laser had been invented, and Bell Laboratory scientists (and others) began exploring its far greater bandwidth possibilities. Because optical radiation has a frequency of more than 10^{14} Hz, the available information bandwidth is very large in comparision with microwaves systems, which have carrier frequencies of about 10^{10} Hz. This offers the attractive possibility of transmitting many more messages over a single carrier network. However, its full practical realization depends upon new techniques for modulating, guiding, and detecting light.[11] A review of the first years of research and problems in this field has been given by Miller and Tillotson.[12] They discussed the problems associated with atmospheric transmission, such as the effects of small amounts of turbulence, which are not important in photography where typical exposures are a significant factor of a second but could be quite serious in communications applications, where integration times as short as 10^{-9} sec are of interest.

As an alternative to atmospheric transmission, the concept of optical *waveguides* has received much attention; these are now referred to generally as optical *fibers*, and we shall see that for short distances, such circuits are already finding commercial applications.

12.10. Basic Laser Communication Considerations

There are a number of developments necessary before even partial utilization of the inherently wide bandwidth of laser circuits can be achieved. This stems from the fact that optical communication systems have problems in efficiency, modulation, transmission, and detection which are similar to those of present communication systems, and, in addition, they have difficulties uniquely optical in nature.

We mentioned one of these latter difficulties in Chapter 7, pointing out why lasers are not used as low-level-signal amplifiers. (One of the key developments which made the original Bell system microwave radio relay circuits possible, was a low-level, low-noise amplifier triode, called the Morton triode, after its developer Jack Morton.) The high noise of a laser, which, as we noted, prevents its use as an amplifier of low-level signals, is due to the fact that the ratio of Einstein's coefficients,[13] which determines the rates of spontaneous and stimulated emission, varies as the cube of the frequency, so that in the visible and near-infrared regions, the rate of stimulated emissions exceeds that of spontaneous emissions only at very high radiation densities. This is not true of the laser's relative the maser because of the much lower frequency of the microwaves used. Nevertheless, there are some (who have been active in the field) who have proposed the "use of a laser amplifier" ahead of the detector (Reference 12, p. 1301) in optical communication circuits.

A second problem results from the fact that all presently available detectors for the visible and near-infrared utilize a direct interaction between the light-wave field and a solid material having dimensions of many wavelengths. This results in the output of such detectors being a function of the amplitude of the light wave in *each tiny* elemental area (of a size approximating the atomic spacing). Accordingly, such detectors are inherently square-law devices.

A third factor results from the fact that all optical *modulators* to date are designed to operate directly on the (light-wave) carrier by a variation of the optical index of refraction of the modulator material.[14] The large modulating voltages required to generate the needed change in index has accordingly led to large modulating-power requirements.

The above difficulties are in part responsible for the more recent trends in long-distance optical communications systems such as that discussed in a Bell Laboratories paper of August 1974[15] which included the statement: "It is easy to show that for optical carrier systems, binary on-off modulation is optimum." We see that such modulation is not affected by the square-law detector problem, and no laser amplifiers are needed to produce maximum (on) or minimum (off) signals. We shall discuss this system, (which uses optical fibers) in more detail later (Figure 12.7). Let us first review some of the atmospheric propagation systems.

12.11. Line-of-Sight Laser Communications

Already several short-range (less than 10 km.), low-powered optical communication systems are operating both in this country and abroad.[16] Thus, scientists at Siemens Research Laboratories have been investigating the possibilities of such a system over the 5.4 km path between the Munich districts of Obersendling and Biesing. The main aim of the work is to investigate the effect of the atmospheric conditions on laser beam communications.[17] A CO_2 laser is used in the investigations (wavelength of 10.6 μm) with an output power of 5 W. Initially, the use of a He–Ne laser was considered, but measurements previously made in Munich confirmed that the invisible infrared beam is considerably less susceptible to atmospheric influence than the visible He–Ne beam. The wavelength emitted by the CO_2 laser falls in a spectral region in which there is a "spectral window" in the atmosphere. Nevertheless, even the CO_2 laser beam is subject to some perturbing effects, including absorption by water vapor and carbon dioxide in the air, light-scattering by minute particles of water and dust, atmospheric turbulence created by side winds, and solar irradiation. All of these effects tend to attenuate the laser beam, render it more diffuse, displace it, and cause its intensity to fluctuate. Despite this, communication via a CO_2 laser beam still is possible even in heavy mist and moderate rain, fog, and snow. At the

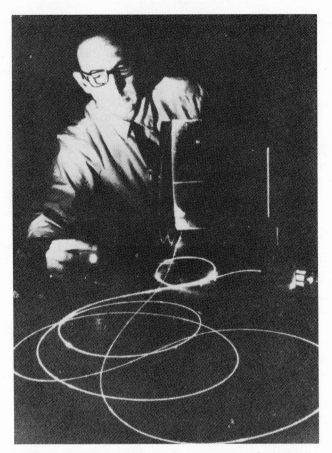

FIGURE 12.7. This experimental fiber for long-distance optical transmission systems is a liquid-filled (tetrachlorethylene) fused-quartz capillary.

present stage of development, the system continues to function satisfactorily until the average atmospheric attenuation of the laser beam exceeds 8 dB/km. The terminal stations are equipped with Cassegrainian telescopes and the laser beam is modulated with a gallium arsenide crystal. The receiving end employs a germanium photodetector.[18]

American Laser Systems and Stabilized Optics recently introduced a lightweight, hand-held laser communications system in demonstrations with the Los Angeles Police Department.[19] Weighing 5 pounds, the unit can reportedly transmit data up to 20 miles, virtually without detection. Possible applications include communications between airplanes and ships, police departments and possibly airports, and within cities for microwave users. The unit includes a hand-held transmitter and a receiver with a gallium arsenide laser diode. Sending about 7000 infrared beam impulses per second, the unit's very low energy levels make it difficult to detect. A separate small unit acts as a modulator to transform voice messages into a digital code that is sent by the laser beam and to decode messages from another transmitter back into voice messages. Los Angeles police considered using the unit in helicopter patrols that send and receive messages from ground patrol cars. The lasers units were to be tested as an aid to overcoming problems with crowded radio frequencies. Used in computer data transmission, the system would not require FCC licenses or the laying of cable.

Optical Communications, Inc. introduced a laser data link (abbreviated LDL) for computer-to-computer communications over short distances, especially in densely populated urban areas. The optical data link supplies an alternative to wire communications.[26] Data transmission of up to 1.544 Mbit/sec (megabits per second) can be reliably maintained over line-of-sight distances up to 1 mile, using a pulsed, infrared gallium arsenide laser diode as the transmitter carrier source. Again, no FCC license or frequency allocation is needed to operate the system. However, the operating range is affected by weather conditions. Infrared signals are converted into electrical pulses by silicon PIN (*P*-type intrinsic, *N*-type silicon) photodiodes, working into an extremely high-gain, low-noise, wide dynamic range preamplifier whose output is coupled into an AGC/limiter amplifier followed by line drivers. The line drivers feed logic circuits for the recovery of timing, framing, and digital data. The LDL system can be operated at any predetermined data rate within the capability of the unit. However, the standard data rates available are in increments of 75×2^n, up to 38.4 kbit/sec, plus 40.8 kbit/sec, 105 kbit/sec, and 240 kbit/sec.

12.12. Satellite Laser Communications

Because synchronous satellites such as the ones in Figure 12.6 cover different portions of the earth, satellite communications transmission between distant points on the globe often require a relaying of signals from one satellite (such as the bottom one of the figure) to another (the left-hand one). At the high altitudes of such satellites, there are no atmospheric effects, which often plague atmospheric laser transmissions, so effort has been applied toward developing *laser* connecting circuits. As the number of stationary satellites continues to increase, the difficulty of accurately aiming microwave beams (so that the beams do not strike satellites other than the desired ones) becomes greater. The sharp beam of a laser avoids this problem. We will now discuss some of the aspects of such systems.

A neodymium–YAG laser capable of transmitting one billion bits of information per second between satellites was announced by GTE-Sylvania for an Air Force laser communication system.[21] With peripheral equipment of less power and weight than present satellite radio networks, the Sylvania laser system can send more data faster and operates on a narrow bandwidth, reducing interference and interception possibilities. Battery-powered light from a small lamp stimulates the YAG material to produce a $\frac{1}{4}$-W power beam capable of generating 500 million pulses per second. Pulses are then coded to provide a data transmission rate of one billion bits per second. The work is based on new techniques developed for stabilizing the beam, and for cooling the laser by conductive cooling (with the heat dispersed to the body of the satellite and radiated into space). In this rather ambitious communication program (commonly referred to as 405B), the U.S. Air Force is developing space-qualifiable hardware to allow data to be relayed from synchronous to synchronous, low-orbit to synchronous, and synchronous to earth, with an error rate of one part in a million.[22]

The demonstrated pointing accuracy, with a stimulated spacecraft environment, is better than one microradian. From this program, technical advances have been achieved in several areas, including quality-modulator material of lithium tantalate, fast avalanche photodiodes with peak sensitivity at 1.06 μm (the fastest known detectors of good quantum efficiency), and fast cross-field photomultipliers, both dynamic and static, with response to 3 GHz.

12.13. Wires of Glass

Twenty-five years ago, fiber optics, the "light-pipe" phenomenon, was virtually unknown. Fifteen years ago it was a laboratory curiosity. Today it is a multimillion-dollar business with thousands of applications. The impact of fiber optics is being felt in many fields, including the automotive industry, where such fibers are being considered for (among other things) indicating failed headlights.[23,24]

Perhaps the largest use could come in the communications field, as newer and better components, including the light pipe itself, becomes available. Because the loss in the glass plays a very important part in such systems (it determines how close together repeater stations must be placed), extensive efforts have been directed toward lowering this loss per unit length. In the routine manufacture of fiber optics, attenuations of 1000 dB/km often result. While this does not affect certain fiber applications, it would obviously be devastating to a communications channel. Advances in both single-mode and multimode coaxial glass fibers resulted, in the early 1970's, in losses as low as 18 dB/km, as compared to a nominal 100 dB/km for better-quality optical fibers. Bell Laboratories has also considered fibers with liquid cores and found losses as low as 14 dB/km at a wavelength of 1.08 μ [25] (Figure 12.7). Earlier low-loss "coaxial" glass fibers were fabricated with two different materials—one for a very narrow inner region called the core, and the other for a surrounding outer cladding. Light in transit through a glass fiber is kept in the core region by the inward-focusing effect in the outer cladding. Until recently, fibers made with differing glass materials often contained undesired impurities that interfered with the passage of light and caused transmission losses.

More recently, Bell Laboratories scientists made it possible to fabricate efficient light-carrying glass fibers from a single material. The new hair-thin fibers are made with the purest known commercially available glass. [26,27] The new fiber has shown light loss as low as 5 dB/km (50% in 2000 ft). This would allow signal amplifiers to be placed further apart than in land cable systems now in service. Still more recently, [28] Bell announced a *low-loss*, low-*dispersion* fiber. In this structure, the smearing of light pulses by dispersion, caused when some light within each pulse travels in the inwardly focused zigzag path during passage, is reduced. This is accomplished by varying the composition of the fiber radially, to produce a *graded* index of refraction, causing the light following the zigzag path to travel faster than the direct light so that all of a given pulse arrives at the destination at almost the same time. New techniques for *making* the fibers have also been developed, [29] one of these involving the use of a CO_2 laser (Figure 12.8 and 12.9). The laser, acting as a highly controllable, clean source of heat, melts a high-purity glass rod, permitting it to be drawn into a hair-thin fiber a mile long.

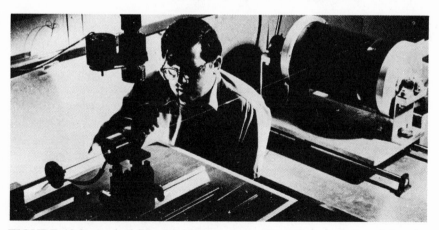

FIGURE 12.8. An invisible CO_2 laser beam melts a glass rod, permitting the glass to be drawn into a fine fiber. The drum at the right holds the drawn fiber (courtesy of Bell Laboratories).

FIGURE 12.9. A close-up of Figure 12.8., showing, at the front left, the fiber useful for transmitting possibly 4000 telephone conversations at once.

12.14. A Long-Distance Fiber Circuit

We mentioned earlier the experimental Bell circuit employing binary pulses.[15] This system makes use of a new, small, room-temperature gallium arsenide laser, which can be powered by a small battery. This laser, the size of a pinhead, operates at room temperature and is a development which first occurred in 1970.[30] Its beam is capable of being modulated at a rate corresponding to the transmission of a billion bits of information per second, equivalent to sending, letter-by-letter, 200 average-size books in one second.[30] Such lasers have operated continuously for over one year. Although another light source, the light emitting diode (LED), can become a competitor to the gallium arsenide laser, two factors favor the laser. The first involves the radiation pattern. The diode emits light uniformly over all angles, whereas a laser produces a directional beam; this provides a 15-dB increase in coupling to a fiber.[15] Secondly, the bandwidth of the light from a diode is typically 360 Å, whereas that from the laser is smaller than 20 Å. Even small amounts of delay distortion (dispersion) inherent in the fiber act unfavorably on the wide bandwidth of the diode.[15]

Another key element in this experimental system is in the receiver portion. It is the silicon photodiode detector; two versions are possible, a PIN photodiode and an avalanche-type photodiode. In both versions, carriers are generated by the photons with high quantum efficiency; in the PIN type, the carriers are collected by a bias voltage across the diode, while in the avalanche type they are amplified by electron–electron interaction, as in secondary emission.

Because of the large reduction in attenuation through the recent developments in new fibers, these "glass-wire" communications circuits hold much promise. Prior to 1970, the attenuation was 20 dB/km, i.e., the emerging power at the 1-km point was 0.01 of the entering power. In 1973, a figure of 4 dB/km (for GaAs-laser wavelengths) was reached, and in 1974 a light loss lower than 2 dB was achieved. This 2 dB/km figure compares with 5 dB/km for coaxial cable at 10 MHz.

In summary, the complete system comprises: encoders for converting the many (analog) speech signals to binary pulses, a time-division multiplexer to put all of the pulse signals in a proper time sequence, a driver to activate the laser when an "on" signal pulse occurs, a low-loss glass fiber, a photodiode receiver to convert the light pulses to current pulses, and an amplifier to build the pulses back up to their original value. At this stage, the process can be repeated using only a driver to activate a new laser at this "repeater station." At the final receiving end, a time-division *demultiplexer* sorts out the individual telephone messages and a *decoder* transforms the binary versions back into analog speech signals.

12.15. Short-Length Fiber Circuits

Fiber-optics circuits are attractive as wiring circuits within aircraft[31] and on shipboard[32] because they avoid electromagnetic interference and reduce the susceptibility of the on-board circuits to gamma rays, lightning, and, for the military, the electromagnetic pulse associated with a nuclear explosion. Such a fiber-optic data link carrying aircraft flight control signals from cockpit to controls was successfully flight tested by the U.S. Air Force as part of a program to evaluate various transmission media for carrying multiplexed signals in a fly-by-wire flight control system. Of particular concern was the potentially catastrophic effect of lightning and other forms of electromagnetic interference on the conventional twisted-pair-wire bus now used to carry primary flight control signals. The two-way multiport fiber-optic data bus was integrated with a digital air-borne data system.[33]

The first known fiber-optic communications network was installed on

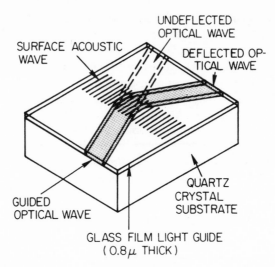

FIGURE 12.10. A surface acoustic wave causes an optical wave traveling in a thin glass film to be deflected (courtesy of IBM).

shipboard on a ship in the U.S. Sixth Fleet. The cable was part of a six-station system built by the Naval Electronics Laboratory Center in San Diego.[32] Manufacturers involved in such short-length systems include Hughes, Galileo Electro-Optics, and Spectronics. Advantages include: total electrical isolation; no dielectric breakdown; no ringing or echoes; a reduction in weight, power requirements, and cost; and a large increase in bandwidth.[31]

12.16. Optical Film Devices

In thin glass *films*, rather than fibers, a technique has been developed for deflecting light waves traveling in the film using surface acoustic waves (Figure 12.10). The waveguide (the thin glass film) is deposited in a quartz crystal substrate and an acoustic surface wave of 200 MHz is propagated on the crystal surface at approximately right angles to the guided optical wave. A deflection efficiency of 66% was obtained.[34]

12.17. Lasers in Microfilm Transmission

Using a pulsed laser beam whose intensity is varied in accordance with the information in a microfilm, images are recorded on frames of a 16-mm transparent mylar film coated with a thin film of indium–tin oxide. The laser burns tiny holes of varying diameter in the coating, producing almost instantly a high-fidelity transparency without a development process. The technique was reported as able to reproduce a newspaper page sent over a high-capacity transmission channel in 4 seconds, an operation taking 4 minutes over a telephone line.[35] A similar system was developed by RCA.[36]

Other lasers transmission systems have been applied in phototypesetting by Photon, Inc.,[37] in sending wirephotos (by Associated Press and MIT),[38] and in newspaper facsimile systems (Laser Graph, Sudbury, Mass., and Matsushita Graphic Communications, Japan).

12.18. Television Recording

A system using a laser as a "stylus" to play back a television program previously recorded on a plastic, microgrooved disk, was developed by MCA Disco-Vision Inc.[39] The grooved disks are capable of storing 40 minutes of television playing time per side. Unlike conventional long-playing record

FIGURE 12.11. A laser color projection system produces the picture shown at the top right.

players, the device uses a noncontact pickup arrangement, a laser beam, to scan the grooves and extract the video-audio information they contain.

Other approaches to TV playback have been announced by N. V. Phillips (also using a transparent disk scanned by a He–Ne laser), and Thomson-CSF (also using a flexible disk and a laser optical reader). In most of these systems, the plastic disks (mylar) are pressed from masters and are expected to be rather inexpensive.

Still another method involves *recording* the information with modulated laser beam on a *photographic* disk. Playback can then be done with an inexpensive miniature incandescent lamp, avoiding the cost, to the home user, of the laser.[40]

A technique described by RCA quite early and called at that time Selectavision involves recording the TV picture information as *holograms* which are embossed on inexpensive vinyl tape and reconstructed for pickup by an electronic converting device, by a laser.[41] At the Fifth Holography School held in Novosibirsk, Siberia in January 1973, in which this author also participated (Figure 14.28), a paper on this system and a demonstration of it was presented by W. J. Hannan, and it drew by far the largest attendance of any paper.[42]

In playback procedures, laser projection systems can be useful (Figure 12.11).

12.19. References

1. W. L. Barrow, Transmission of electromagnetic waves in hollow tubes of metal, *Proc. IRE*, **24**, Oct. 1298 (1936).
2. J. H. Vogelman, Microwave communications, *Proc. IRE* **50**, 907–911 (1962).
3. W. E. Kock, *Radar, Sonar and Holography*, Academic Press, New York, 1973.
4. H. T. Friis, Microwave repeater research, *Bell Syst. Tech. J.* **27**, 183–246 (1948 (in this composite report, this author's section, "Antenna Research," describes the lenses of Figure 3.7 and 10.3).
5. W. E. Kock, *Sound Waves and Light Waves*, Doubleday, Garden City, N.Y., 1965, p. 128.
6. A. C. Clarke, Extra-terrestrial relays, *Wireless World*, Oct., 305–308 (1945).
7. W. E. Kock, Physics in the missile and satellite fields, *Phys. Today*, **13**, 24–29 (1960).
8. A. H. Waynick, Satellite area communications including the USSR, European Scientific Notes, ORNL, July 31, 1974, pp. 242–245.
9. W. E. Kock, Communications in the space age, *Bridge Eta Kappa Nu*, **64**, (Feb.) 18–19 (1968).
10. *Opt. Spectra* **8**, Aug., 10 (1974).
11. R. D. Haun, Jr., Laser applications, *IEEE Spectrum* **5**, May, 92 (1968).
12. S. E. Miller and L. C. T. Tillotson, Optical transmission research, *Proc. IEEE* **54** (10), 1300–1311 (1966).

13. B. Lengyel, *Lasers*, Wiley, New York, 1971, p. 178.
14. G. Kane, What's ahead for optical communications?, *Opt. Spectra* **6**, Oct., 24–38 (1972).
15. M. DiDomenico, Jr., Wires of glass, *Ind. Res.* **16** (9), 50–54 (1974) (presented at the 1974 IEEE Solid-State Circuits Conference, Philadelphia, Pa.).
16. *Ind. Res.* **14** (5), 19 (1972).
17. *Lasersphere* **2**, July 15, 4 (1972).
18. *Eurosci. Intell. Rep.* **July**, 4 (1972).
19. *Electron. News* **18**, May 7, 24 (1973).
20. *IEEE Spectrum* **10**, Aug., 70 (1973).
21. *Opt. Spectra* **7**, Apr., 14 (1973).
22. *Laser Focus* **9**, Nov., 6 (1973).
23. *Electron. Prod.* **June**, 152 (1968).
24. J. Kloots, Putting fiber optics to work, *Opt. Spectra* **5**, June, 22–34 (1971).
25. T. Li and E. T. Marcatili, Research on optical fiber transmission, *Bell Lab. Rec.* **49**, Dec., (1971).
26. *IEEE Spectrum* **10**, July, 85 (1973).
27. *Bell Lab. Rec.* **51**, July/Aug., 220 (1973).
28. *Electron. News* **19**, June 24, 31 (1974).
29. *Opt. Spectra* **7**, Nov., 40 (1973).
30. *Lasersphere* **3**, June/July, 2 (1973).
31. J. N. Kessler, Fiber optics excites the military, *Electronics* **47**, Aug. 22, 69–71 (1974).
32. *Laser Focus* **10**, Feb., 74 (1974).
33. *Electronics* **47**, Aug. 8, (1974).
34. *Microwaves* **10**, Mar., 26 (1971).
35. *Bell Lab. Rec.* **50**, Nov., 62 (1972).
36. *Ind. Res.* **14** (11), 23 (1972).
37. *Lasersphere* **2**, Dec. 15, 1 (1972).
38. *Lasersphere* **3**, May 15, 1 (1973).
39. *Lasersphere* **2**, Jan. 15, 1 (1972).
40. *Electronics* **47**, Apr. 4, 114–118 (1974).
41. J. B. Scott Lasers and Holography, a new approach for a television tape player, *Microwaves* **8**, Dec., 19–24 (1969).
42. W. E. Kock and G. W. Stroke, International communication: fifth holography school, Novosibirsk, 29 January–3 February, 1973, *Appl. Opt.* **13**, Nov. A14–A17 (1974).

13

MEDICAL
APPLICATIONS

We shall see that both fields, those of lasers and of holography, have found wide uses in medicine. Promising uses of lasers are found in such fields as ophthalmology (cataracts, retinal detachments), dentistry (enamel glazing, tooth contouring), and laser-beam surgery. However, it would appear that the most signficant medical applications are now found in the field of holography, with acoustic holographic techniques being in the forefront because of the earlier extensive use of conventional (sonar) ultrasonic imaging (a form of acoustic "x rays") in medical diagnostics. We shall therefore review first the holographic applications.

13.1 The Development of Ultrasonics in Medical Imaging

One of the first things that happened to the visitor to the laboratory of Dr. F. A. Firestone at the University of Michigan back in 1944 was to be invited to stick his foot into a bucket of water.[1] "I will show you the bones," said the inventor of a device that was then used to detect flaws in metals by the reflection of high-frequency sound waves.[2] Although Firestone's device was not immediately used in medical applications, its capabilities for penetrating the human body and returning echoes from various tissue structures without the accompanying hazards of x radiation soon attracted medical researchers. In such systems a transducer is connected with a signal source and coupled, through a suitable transmission medium, to the subject. Because the transducer can both send out and receive ultrasonic energy, the display of

313

the modifications that its energy undergoes as it is transmitted, refracted, or reflected by different bodily tissues makes it acquire information about the size, location, and character of interior structures. The interpretation of this information constitutes the present art of ultrasonic diagnosis.

A review of the recent literature on ultrasonic diagnosis reveals impressive gains in clinical situations, and also how much of an art it still is. It takes an experienced physician to interpret correctly the anatomical features called out in the two-dimensional ultrasonic visualization.[1] Such "obviously recognizable echograms," writes Dr. George Kossoff of the Commonwealth Acoustic Laboratory in Sydney, Australia, "are useful for interpretive experience. For instance,... until clear spine echoes were obtained, it was difficult to distinguish a fetal trunk from a fetal head, both of which gave a circular cross section. Once the spine echoes had been identified, the trunk and head could be recognized in less clear echograms. Similarly, various other structures such as the placenta, the fetal heart, and fetal kidneys have now been recognized."[3]

13.2. Recent Developments

Medical ultrasonic diagnostics has recently become a field of major activity in clinical applications as well as in industrial exploitation. Productivity in health care delivery and early screening for possible disease may indeed be considerably improved by use of the new methods of ultrasonic diagnostics which are now under intensive development in leading academic, national, and industrial laboratories around the world. Advanced prototype ultrasonic imaging already makes it possible to reveal noninvasively (in an x ray-like manner) disease (or the absence of it) in almost all human, internal soft-tissue organs including the prostate, the bladder, the heart, and the breasts, among others. A widespread use of ultrasonic imaging is also made in obstetrics. Strikingly remarkable images of fetuses are being readily obtained with needle-like sharpness in the best cases, starting even earlier than the 10th week of pregnancy. The method permits one to guide an aspirating needle in order to extract some of the amniotic fluid without injury to the fetus, with a view toward examining chromosomes for possible abnormalities. These are just a few representative illustrations of the many diverse uses of the method of medical ultrasonic diagnostics which are now being perfected.

The importance of this field is indicated by the reports of recent clinical experiments which have demonstrated the usefulness of ultrasonic imaging for diagnostic visualization. The method is complementary to radiography, nuclear medicine, and thermography, among others. It may be used safely, notably in cases where other methods are either inapplicable or perhaps hazardous (e.g., as are ionizing radiations in the case of pregnancies). Many of the recent advances in ultrasonic diagnostics were reported at the Third U.S.–Japan Seminar[4] on Imaging and Holography held in Hawaii in 1973 (Figure 13.1), and sponsored by the National Science Foundation and the Japanese Society for the Promotion of Science.[5] The wide range of developments disclosed at that seminar prompted the National Science Foundation to form a special "Blue Ribbon Task Force on Ultrasonic Imaging" in February 1973. Under the chairmanship of John B. Manniello, a special consultant to NSF (then Vice President of Government Operations at CBS Laboratories, and more recently, Science Attache to the U.S. Ambassador to Italy) and under the direction of C. Branson Smith, Director of the Office of Experimental R & D Incentives at NSF, the Task Force carried out a world survey of the state of the art in ultrasonic medical diagnostics and made recommendations for accelerating more widespread clinical use of ultrasonic instrumentation and increased industrial participation. The Task Force consulted with leading medical and industrial authorities in the world, in countries ranging from the United States, Japan, Germany, Great Britain, and Austria, to Denmark,

FIGURE 13.1. Photographs taken at the Third U.S.–Japan Seminar: (Top, left to right) Professor Dennis Gabor, Nobel Laureate, Japanese Co-chairmen, Y. Kikuchi and J. Tsujiuchi, and (Bottom) U.S. Co-chairmen G. W. Stroke (left) and W. E. Kock.

Sweden, Holland, and Australia, among others. It reported through C. Branson Smith to Dr. H. Guyford Stever, Director of the National Science Foundation (concurrently also, the Science Advisor to the President of the United States). It included among others, Professor Dennis Gabor, Nobel Laureate, and Dr. Frederick Seitz, President of Rockefeller University. An official report of the Task Force was published by the National Science Foundation.[6]

13.3. Details of Recent Ultrasonic Techniques

As noted above, the use, within the last decade, of ultrasonic diagnostics for various portions of the human body has grown rapidly. Those body areas include the neck (in searching for cancer in the thyroid), the head (for brain tumors), the female breast (for cysts and tumors), and the abdomen (particularly of pregnant women for examination of the growing offspring). The most common scan procedure for such ultrasonic diagnosis equipment is called the B-scan;[7] it provides a picture of a cross section of the body portion scanned, the picture being referred to as a tomograph. Typically, these sonarlike devices employ an ultrasonic transducer having an effective aperture 1–2 cm in diameter, and an ultrasonic frequency of 2 MHz (these waves having a wavelength of 0.75 mm in water). The (pulsed) transducer is placed in contact with the skin by the use of a fluid or a gel (to minimize the impedance mismatch should air be present between the transducer and the skin), and a linear scan is made across the area of the body where a section (tomograph) is desired. The received acoustic echoes are displayed on a cathode ray tube, with the horizontal plot (abscissa) following the lateral movement of the transducer, and the range (the distance to, or depth of, reflecting areas within the body section) being plotted vertically. The intensity (the brightness of the cathode-ray-tube spot) is made proportional to the strength of the echo.

Because of the short wavelengths used, the near field of the transducer extends out many inches, so the beam, both when transmitting and when receiving, maintains a width equal to the transducer diameter. Individual objects (such as cysts) which are smaller than this dimension, and which are in groups, cannot be resolved. Furthermore, as seen in Figures 1.14 and 13.2, near-field patterns are quite nonuniform,[8] causing small objects to generate reflections which fluctuate in intensity as the beam is scanned. This problem of near-field intensity variation in B-scan devices has been discussed by Kikuchi.[9] One procedure for alleviating this problem to some degree involves the use of a focused transducer. Kikuchi has recommended that the quantity $R^2/a\lambda$, where R is the radius of the transducer, a the range of the focal point, and λ the wavelength, should be set at about 4. This value provides a compromise situation, with a fairly smooth *axial* variation along the beam and a fair sharpness in the focal region.[9] The focused area is comparable to that formed by the inwardly focused beam of the lens of Figure 10.1 (the uniform, focused white area in the figure). The various amounts of *fixed* inward focusing which have been recommended for such transducers certainly do improve the performance, but maximum improvement occurs only at the focal point (as in Figure 10.1).

Kossoff has described the use of *several* elemental transducers (in place of the single one) in a B-scan system,[10] with a *variable* focusing procedure then being used, but only in the receiving process. In this variation, the outgoing pulse still travels along the wide, near-field beam (Figure 13.2), and, in addition, the maximum receiving focusing is only modestly approximated. We saw, in connection with Figure 10.13, that to truly achieve maximum inward focusing effectiveness in near-field sonar-type systems a large number of transducers are required, and a very complicated, time-varying phasing process is needed to achieve the varying focusing effect. In addition, the wavefront focusing process must be rapid enough to cause the focused region to follow the pulse as it moves outward from the transducer array. Even when this complicated operation is carried out, focusing is only achieved along the axis of the transducer array.

13.4. Synthetic-Aperture Techniques in Medicine

Obviously, if there were some magic way of causing the focused beam to always have an ideal sharpness *at all points along the beam* this would be the far superior choice. If, in addition, this procedure could cause the beam to be optimally focused at *off-axis* points in the near field, it would be even more attractive. As we saw above, this is exactly what the synthetic-aperture technique accomplishes automatically, furnishing, in addition, a focal spot

FIGURE 13.2. The intensity variation in the near field of an acoustic radiator.

diameter equal to that of a focused transducer of much larger dimension (a *synthetic* aperture) than that of the actual transducer. We saw in Figure 9.27 one particularly prominent focusing zone plate in the myriads of super-imposed zone plates in that figure, and, in the photographs which followed that figure, the highly detailed "tomographs" of the *terrain* flown over by the aircraft! For those photographs are in truth two-dimensional plots of echoes, just as are the two-dimensional B-scan records (tomographs or plane *sections*) of living tissue.

This excellent, near-field focusing property of holograms and synthetic-aperture systems makes that development an excellent candidate for medical ultrasonic diagnostics.[11] We noted earlier in connection with Figures 10.8 and 10.9 that this use was described in a 1970 U.S. Patent (see Chapter 9, Reference 39). More recently, the synthetic-aperture technique has been applied directly to the B-scan process. In these experiments,[12] the vertical dimension (aperture) of the B-scan transducer was not changed, and instead of making the horizontal aperture small (as is customary in synthetic-aperture radar or sonar systems so as to provide a wide angular beam), a diverging lens was placed in front of the transducer to achieve a similar effect. This system[12] achieved, for closely spaced wires of a wire grid (corresponding to small reflecting cysts in medical diagnostic), resolutions which were ten times better than the standard, equivalent, B-scan-equipment resolutions.

13.5. Real-Time Synthetic Aperture

One disadvantage noted in the discussion of these experiments[12] is that in usual holographic or synthetic-aperture systems a photographic recording and reconstruction process is involved, incurring a delay in the examination of the resulting records. More recently, it was pointed out[13] that the real-time Goetz tube (Figure 9.6) could be useful in permitting the hologram technique to provide a viewer with an immediate, well-detailed, section view (tomograph) of the desired body area. Figures 13.3 and 13.4 indicate the information-handling capability of an early model of this tube, the tube often being referred to as an optical computer.

13.6. Liquid-Surface Holography for Medical Diagnostics

We noted above[7,9] that extensive development has occurred in the "sonar-type" ultrasonic imaging field, and that much attention has been given to those techniques in which pulse–echo methods have been used to generate

FIGURE 13.3. The record of two sine waves established on the DKDP crystal of the Goetz real-time optical computer tube of Figures 9.6 and 9.7.

B-scan displays, among others. Accordingly, many in the medical field are familiar with one or more of these systems. Ultrasonic holography, and, in particular, the practical acoustical imaging systems based upon liquid-surface acoustical holography, are not, as yet, so familiar to their potential users.

Although work has been done on various aspects of scanned acoustic holography, and many interesting images have been obtained by scanning methods, the most rapid development has been in the liquid-surface technique (which we discussed briefly in Chapter 4 in connection with Figure 4.13), and

FIGURE 13.4. Reconstruction of the image of the "hologram" of Figure 13.3. The two sine waves of 450 and 500 kHz are reconstructed as real and virtual images (the pairs of dots on the left and right) (Bendix photographs).

such equipment is now available commercially. Figure 13.5 shows such an instrument. As noted, there are advantages in using a liquid surface over photographic film as a hologram detector: (1) no development is required, and (2) the liquid surface responds rapidly to the ultrasonic energy, so that an instantaneous readout of the hologram is possible.

Normally, no lenses are used in holographic imaging. (Focused-image holography which we discussed in connection with Figure 6.13 is an exception.) Each object point generates its own zone plate and the hologram is really a complex collection of these zone plates, with each zone plate acting as a lens to image its corresponding object point. In the equipment of Figure 13.5, however, an acoustic lens is used to image the object information on the liquid-surface hologram as shown in Figure 13.6. The interaction of sound with the liquid surface is such that, where the object has good transmission, a strong ripple pattern will develop on the liquid surface due to the interference of the object information with the reference-beam wave. Where the ripple pattern is strong, much light will be diffracted out of the main beam; on the other hand, where the transmission through the object is poor, the ripple pattern developed on the liquid surface will be weak and there will be less light diffracted out of the main beam. Figures 13.7 and 13.8 show the use of this equipment in nondestructive testing.

It should be noted that the object is back illuminated and imaged by

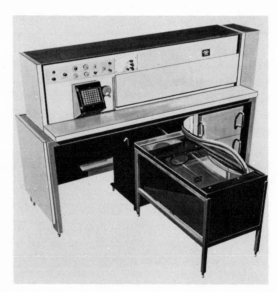

FIGURE 13.5. A liquid-surface holographic real-time imager for medical diagnostics (and for nondestructive testing) (courtesy of Holosonics, Inc.).

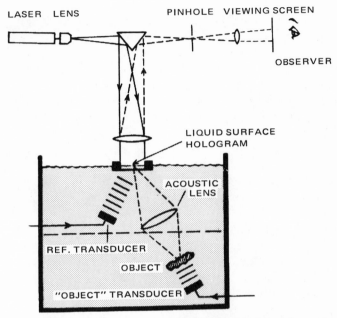

FIGURE 13.6. A schematic drawing of an acoustical holography system.

through transmission (Figure 13.6). The resulting images are therefore subject to different interpretation from those resulting from a B-mode scan, in which the primary interaction mechanism is reflection. Images produced by through-transmission are primarily the result of the absorption characteristics of the object, although scattering and reflection play important roles also. This difference in interaction mechanism is quite apparent in comparing images of hard tumors with liquid-filled cysts in the breast. Figure 13.9 is a typical image of a fluid-filled cyst; this cyst was also detectable by manual palpation. In contrast with the image of the cyst in Figure 13.9, the tumor of Figure 13.10 is shown to be highly absorbing. Both of these pictures were taken using an ultrasonic frequency of 3 MHz.

The ability to diagnose small breast tumors and cysts may prove to be the greatest service of acoustic holography. This capability was demonstrated in 1969,[14] and the prospects that extremely small tumors (as in the female breast) will be detectable appear excellent. Weiss[15] has pointed out that small lumps (1 mm to 1 cm in diameter) are not generally susceptible to detection by manual palpation. Yet if cancerous lumps are not diagnosed at their early stages, the cancer cells spread to other parts of the body to form secondary growths not in contact with the primary cancer; this generally develops into "advanced cancer," which is virtually incurable.

In comparing through-transmission acoustic images with x rays, the first

FIGURE 13.7. A nonbond detected acoustically in an aluminum aerospace structural honeycomb (the central black area).

major difference is that the acoustic image is normally presented as a positive whereas the x-ray image is presented as a negative. Bones, being both highly reflective and highly absorptive to ultrasound appear dark on the acoustic image but light on the x-ray image. In Figure 13.11, the acoustic image is portrayed as a negative, rather than the usual positive,[16] so that in both of these photos the bones are white. In liquid-surface holography the various types of soft-tissue structure are revealed.[17] An acoustical image of the hand at 3 MHz is shown in Figure 13.12; it clearly shows the flexor tendons used to manipulate the thumb, as well as other tendons and the vascular structure. Figure 13.13 shows an acoustical image of the forearm, showing the radial artery. Large veins and arteries are particularly amenable to imaging by ultrasound at 3 MHz. The image shows both sides of the blood vessel as dark lines, with the center of the blood vessel appearing clear or transparent. Imaging of excised organs often provide corroboration to a physician's

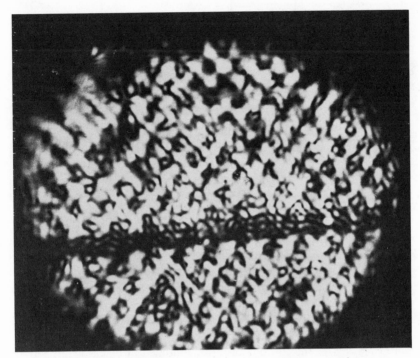

FIGURE 13.8. A line defect in a woven fiber composite.

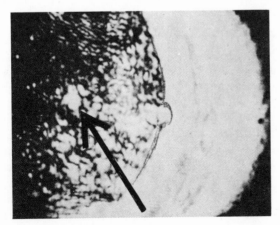

FIGURE 13.9. The arrow points to a fluid-filled cyst (the white area) in a female breast (courtesy of M. D. Anderson Hospital).

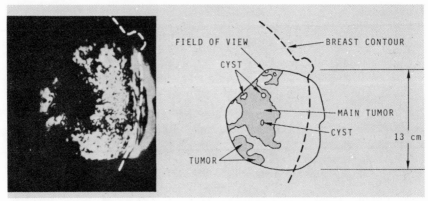

FIGURE 13.10. A breast *tumor* is highly absorbing; it is the black area at the left of the photograph (courtesy of M. D. Anderson Hospital).

diagnosis. Thus, the hologram image (at 5 MHz) of an excised gall bladder (Figure 13.14) shows a smooth, thin wall near a single stone of pure cholesterol crystal.[18] Figure 13.15 shows details of a stillborn human fetus.

For diagnosing many parts of the human body, such as the hands, elbows, knees, feet, etc., the arrangement of Figure 13.5 is quite satisfactory. Thus, Figure 13.12 was obtained by having the person immerse his hand in the black, liquid-filled structure in the foreground, with the reconstructed hologram appearing on the cathode ray tube (outlined in black) at the left side of the equipment. For the ultrasonic holographic diagnosis of the female breast, the arrangement shown in Figure 13.16 is employed, with the breast then submerged in the liquid.

13.7. Concern About Acoustic Intensity in Holography

Some who have worked with conventional B-scan techniques have assumed that the generation of a liquid-surface ripple pattern automatically involves such high sound-energy levels that they could be dangerous to human tissue. This fear was recently allayed by Weiss,[15] who noted that peak intensities of 2.5 W/cm^2 caused no damage, and that normal (Holosonic) intensities were only 0.004 W/cm^2. Similarly, in B-scan systems, a total exposure of 150 W-sec/cm^2 produced no damage, whereas the normal *holographic* exposure is only 1.2 W-sec/cm^2. Weiss also stated that "a search of the literature has failed to reveal evidence of tissue damage caused by exposure to as little as 100 W-sec/cm^2 at any frequency of ultrasonic irradiation." Weiss was particularly concerned because of the "background of X-irradiation-induced cancer." [15]

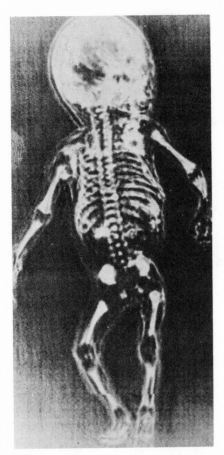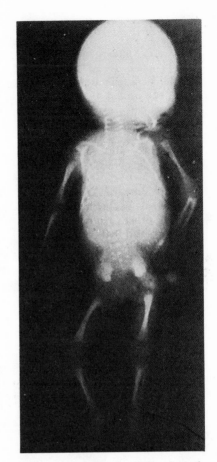

FIGURE 13.11. A comparison of x-ray (right) and ultrasonic (left) *through-transmission* images. The acoustic picture (a negative reproduction) shows more details of a 16-week fetus.[16]

13.8. Holography and Three-Dimensional X-Ray Pictures

An x-ray procedure has been described in which eight to ten weakly exposed x-ray-film records are made of a subject, with the direction of the x-ray beam being changed slightly after each film is exposed. The individual films reveal very little detail because of the weak exposure but when all are superimposed properly, images of details at various depth planes become visible.[19] Because of the successive angular changes used in

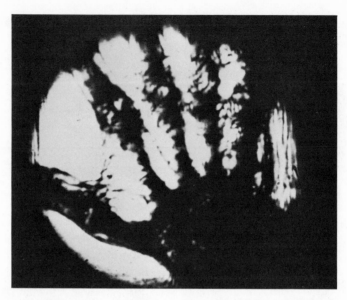

FIGURE 13.12. Acoustic holographic image of a human hand (courtesy of Holosonics, Inc.).

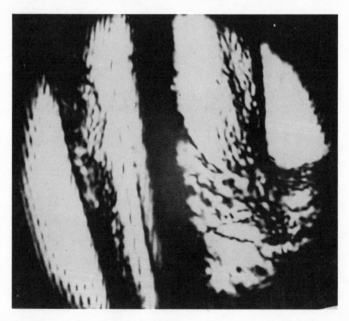

FIGURE 13.13. An ultrasonic image of the arm just above the elbow (courtesy of Holosonics, Inc.).

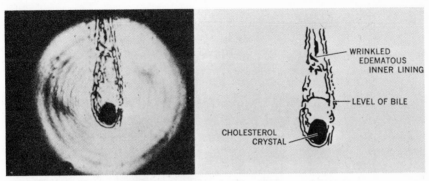

FIGURE 13.14. Reconstructed hologram of a gall bladder showing a stone (of cholesterol) (courtesy of Holosonics, Inc.).

the exposures, portions which are located deeply within the subject require a different arrangement of stacking of the films to become visible than do those near the surface. Accordingly, the particular stacking arrangement which is needed to make an object most clearly visible is a direct indication of the depth of location of that object within the subject which was x-rayed. For example, the depth position of a bullet within a person's body can be ascertained, so that the safest and most effective surgery can be selected for its removal. Presently, a mechanical shifting device is employed which varies the stacking arrangement of the x-ray films in small, successive steps, thereby revealing, at various stacking positions, objects at various location depths.

Quite early in the development of laser holography it was shown that many exposures of the same film could be made, thus generating a *composite* hologram.[20] Recently a procedure was suggested for making a composite hologram of the (laser-illuminated) stacks of x-ray films in various stages of stacking. Thus, at each step, the superimposed films are laser illuminated, so as to expose, to a hologram plate (properly furnished with a reference beam), the picture visually observed at that step. For the next step in the shifting of the films (which would correspond to making visible any objects located a new plane, displaced in depth, say, a distance Δd), the separation of the plane of hologram plate and that of the cluster of superimposed films is increased by that same distance Δd, and a new holographic exposure is made on the hologram plate. Upon completion of all the successive film shifts and the corresponding increases in the separation of the two planes, the hologram is developed and the recorded images reconstructed. The usual three-dimensional virtual image results, with those reflecting objects which are at great depth within the subject x-rayed being seen behind those at shallow depth. The important parallax effect present in holograms is preserved,

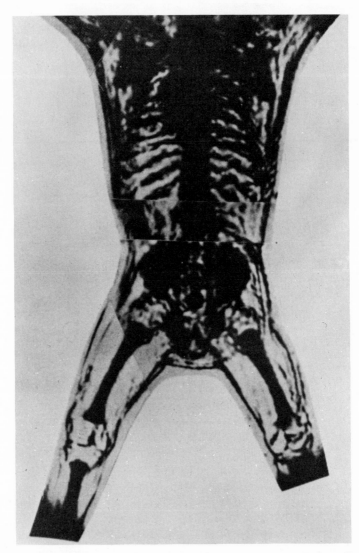

FIGURE 13.15. Holographic image of a 26-week, stillborn human fetus at 5 MHz.

permitting the viewer to look around an object in its foreground by moving his head to the right or left. This ability to view all of the recorded x-ray information simultaneously, although theoretically providing no new information to the viewer, can often assist him in making a better judgment on the situation, particularly through use of the parallax effect.[21]

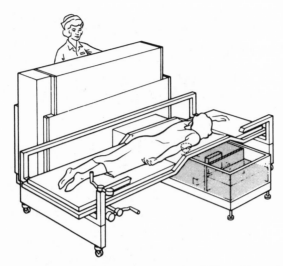

FIGURE 13.16. Procedure for diagnosing the possibility of breast tumors (courtesy of Holosonics, Inc.).

13.9. Holography and Three-Dimensional Gamma-Ray Images

One field of nuclear medicine involves the use of short-lived tracer radioisotopes for diagnosis. Thus, gallium-67 citrate collects in body areas where malignant melanoma, prostatic cancer, or lung cancer exists. The radiation from the gallium-67 in these areas can be made to form an image, permitting these cancerous areas to be localized. The technique is often referred to as radionuclide medical diagnosis.[22] Many of the radioisotopes radiate gamma rays, so that the imaging process is somewhat complicated. The image-forming devices currently in use are various forms of opaque shields having small apertures, including a single-pinhole shield, and a multiple-hole collimator. The detector, placed behind the shield, absorbs the gamma-ray radiation and converts it to a detectable signal, permitting the spatial coordinates of the source to be ascertained.

Barrett[23] proposed the application to nuclear medicine of a technique first suggested by Mertz and Young for imaging x-ray stars.[24] In this process, a lead zone plate is placed between the patient and the detection area, permitting the shadow picture of the zone plate, as generated by the gamma rays radiated from the isotope, to be recorded, through the use of a sodium iodide scintillator plate, on photographic film (Figure 13.17). If the isotope has

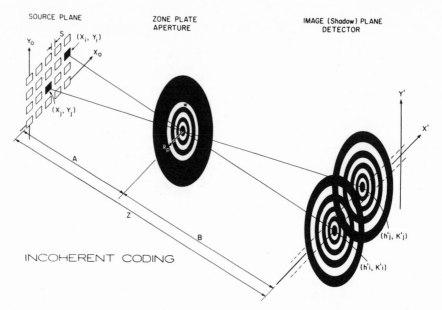

FIGURE 13.17. The two gamma-ray sources (black) at the left cast two shadows, at the right, of the lead zone plate at the center (courtesy of the University of Michigan).

collected in *several* areas of the body (several tumors or cancerous areas), the photograph will record *several* displaced and superimposed shadow pictures of the zone plate. Figure 13.17 shows two sources and two zone-plate records.[25] When the photographic record is drastically reduced in size, the shadow patterns can act as optical zone plates, so that the illumination of the photograph by coherent light (such as that of a laser) results in a concentration of light at the focal point of each of the recorded zone plate shadow pictures. The location of the cancerous areas can thus be ascertained with fairly high accuracy, since the resolution available is determined by the width of the narrowest zone-plate ring (i.e., the outermost ring). Figure 13.18 shows one model of a unit developed at the University of Michigan, or it has an image intensifier positioned behind the lead zone plate at the left.[25]

Figure 13.19 shows an off-axis zone plate. This avoids the superposition of the two hologram images which occurs in an in-line hologram; the desired real image has a background due to the defocused second image. With the *offset* zone plate, the light corresponding to the second image is diffracted in the opposite sense, so neither it nor the zero-order component contributes background to the real image. This effect is seen in Figure 13.15, where the *slightly* off-axis situation has caused a modest separation of the real and virtual images and the zero-order diffraction.

FIGURE 13.18. A nuclear imaging camera; the lead zone plate has an image intensifier behind it.

Figure 13.20 shows that large numbers of superimposed zone plates can reconstruct large numbers of sources. Numerous contributions to this field have been made by the Raytheon Research Division at Waltham, Mass. and by a group at the University of Michigan under the direction of Dr. W. H. Beierwaltes.

13.10. Applications in Ophthalmic Fields

For diagnosing glaucoma, the three-dimensional reconstructions of optical holography are superior to the two-dimensional pictures obtained with x rays, and they permit the exact shape of the back of the eye to be determined.[26] Experiments at the University of Michigan have demonstrated reconstruction with very large depths of field, showing the optic disk and retinal vessels.[27]

Pulsed and continuous-wave argon lasers (in the blue-green region) are focused into a small spot and used for reattaching detached retinas and for performing various other cauterizing-surgery techniques in the eye.[28] In the repair of a detached retina (one that has fallen away from the eyeball), scar tissue is caused to form at several points along the torn retina so that the retina will re-establish close contact with the choroid. Using the laser, no anesthetic is required and negligible pain is sensed during and after irradiation.

An argon laser has been used by an ophthalmic surgeon at Johns Hopkins Hospital to treat retinopathy, an eye disease suffered by long-standing diabetics. The disease involves the weakening of tiny blood vessels in the retina which, when they finally break down, release blood that can spread inside the eyeball and cloud vision temporarily. Also, the blood often causes scar tissue to accumulate on the retina, contracting it and leading to permanent visual impairment.

The argon's blue-green beam can be focused down to a diameter as small as the thinnest retinal blood vessels. The blue-green beam seeks red wavelengths selectively, making the argon laser particularly suitable for work involving blood vessels.[29]

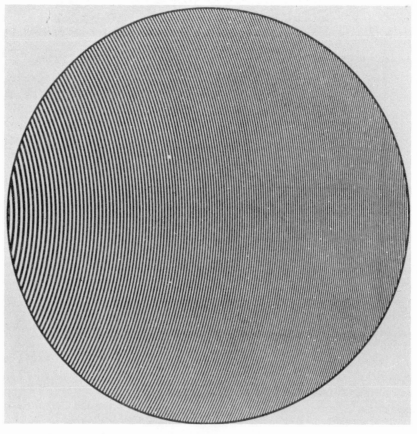

FIGURE 13.19. A zone plate with an off-axis aperture.

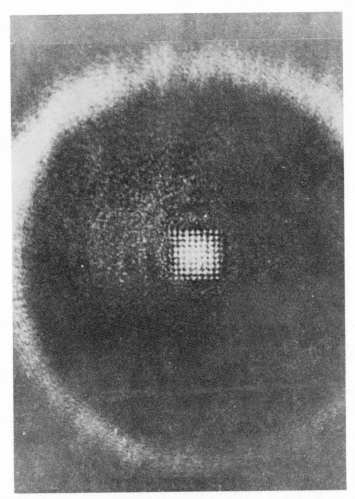

FIGURE 13.20. Reconstruction of large numbers of sources by multiple (super-imposed) zone plates (courtesy of the University of Michigan).

13.11. Dental Applications

The CO_2 laser has been used in dentistry, with the beam being focused to form a wide spot on the tooth. The heat generated causes a fusing of minute cracks which can be starting places for tooth decay.[30] The CO_2 laser is used because the enamel surface of a tooth absorbs energy very efficiently at the laser wavelength of 10.6 μm. One of the pioneers in this field, Dr. R. H.

Stern of the University of California at Los Angeles has noted that enamel decay is caused by demineralization of teeth via organic pathways through the enamel. While lasing has no visible effect on a tooth, microscopic examination indicates that it glazes the outer surface, sealing off these pathways. According to Stern, "this occurs naturally as people get older, which is why old people don't normally have enamel decay, so, in effect, what the laser does is age teeth clinically." [31,32]

The laser is also used for the welding and soldering of crowns and dental bridgework. Rather early use showed that such welds hold up without failure for very long times. [33] More recently, a dental welder was placed on the market [34] which employs a flexible optical fiber. It can thus weld bridgework in the patient's mouth without the delay of deliveries to and from dental laboratories. The delivery system was developed by H. Nath of the University of Munich, [35] and it makes use of the pulsed YAG laser.

In 1974, the U.S. Air Force solicited bids for a prototype of a dental system including a laser for the contouring of teeth and a holography system for producing contouring holograms (for quantitative mappings of topographic characteristics) of the oral structures of living persons. [36]

The laser was found useful in surgery at a quite early point, [37] and Bell Laboratories developed an articulated arm that makes it easier for surgeons to use a laser beam as a scalpel. The beam is guided through several hollow-tube sections that are connected at right angles by hollow blocks. Prisms within the blocks bend the beam 90°. At the 1972 Electro-Optics Conference and Exhibition in New York City a joint paper by two University of Cincinnati and two University of Munich scientists [38] indicated that the high-power output of Nd–YAG systems (up to 200 W) and a new, special, fiber-optics system (which provides high transmission and flexibility) have made it possible to do controlled clinical investigative laser surgical studies. Details of recent instrumentation development and experimentation, including safety programs and initial investigative studies of high-power-output ND–YAG laser surgery in animals and human patients, were described. [39] Laser surgery has also been under investigation at the University of Michigan, [40] where M. W. Berns uses a highly focused argon ion laser beam as a "light scalpel" to cut and remove pieces of chromosomes in an effort to learn more about the function and behavior of cells.

13.12. References

1. N. Lindgren, Ultrasonics in medicine, *IEEE Spectrum* **6**, Nov., 48–57 (1969) (this review article is highly recommended to the reader; it covers mostly the present state of ultrasonic imaging, but it also includes a section on early acoustic holography).

2. F. A. Firestone, U.S. patent 2,2,80,226, Apr. 1, 1942.
3. G. Kossoff, D. E. Robinson, and W. J. Garrett, Ultrasonic two-dimensional visualization for medical diagnosis, *J. Acoust. Soc. Am.* **44**, 1310–1318 (1968).
4. Third U.S.–Japan Seminar, Honolulu, 7–13 January, 1973 *Appl. Opt.* **12**, June, 1068 (1973).
5. G. W. Stroke, W. E. Kock, Y. Kikuchi, J. Tsujiuchi, eds., Ultrasonic imaging and holography in medical, sonar and optical applications, Plenum Press, New York, 1974 (a compilation of the 20 papers presented at the 1973 U.S.–Japan Seminar on the subject).
6. Prospectives for ultrasonic imaging in medical diagnosis, National Science Foundation Publication Nov. 1973.
7. P. N. T. Wells, *Physical Principles of Ultrasonic Diagnosis*, Academic Press, New York, 1969.
8. W. E. Kock, *Seeing Sound*, Wiley, New York, 1971, p. 33.
9. Y. Kikuchi, Present aspects of "Ultrasonotomography" for medical diagnostics, in: *Ultrasonic Imaging and Holography*, Plenum Press, New York, 1974, pp. 229–286.
10. G. Kossoff, Presentation at the NATO Advanced Study Institute on Ultrasonic Medical Diagnostics, Milan, Italy (Professor E. Camatini) June 1974, papers to be published by Plenum Press (1975).
11. W. E. Kock, Synthetic aperture and liquid surface ultrasonic holography for medical diagnostics, NATO Advanced Study Institute, June 1974, papers to appear in book form published by Plenum Press (1975).
12. C. B. Burckhardt, P. Grandchamp, and H. Hoffmann, An experimental 2 MHz synthetic aperture system intended for medical use, *IEEE Trans. Sonics Ultrason.* **SU-21**, Jan., 1–6 (1974).
13. W. E. Kock, A Real-time parallel optical processing technique, *IEEE Trans. Comput.* **C-24**, April, 407–411 (1975).
14. L. Weiss and E. D. Holyoke, Detection of tumors in soft tissues by ultrasonic holography, *Surg. Gynecol. Obstet.* **May**, 953–962 (1969).
15. L. Weiss, Detection of breast cancer by ultrasonic holography, in: *Ultrasonic Imaging and Holography*, Plenum Press, New York, 1974, pp. 567–585.
16. D. R. Hollbrooke, E. E. McCurry, and V. Richards, Medical uses of acoustical holography, in: *Acoustical Holography*, Plenum Press, New York, 1974, Vol. 5, pp. 415–451.
17. R. E. Anderson, Potential medical applications for ultrasonic holography, in: *Acoustical Holography*, Plenum Press, New York, 1974, Vol. 5, pp. 505–513.
18. B. Brendon, Ultrasonic holography: A practical system, in: *Ultrasonic Imaging and Holography*, Plenum Press, New York, 1974, pp. 87–104.
19. A. G. Richards, Variable depth laminography, *Biomed. Sci. Instrum.* **6**, 194–199, (1969).
20. G. W. Stroke, F. H. Westervelt, and R. G. Zech, Holographic synthesis of computer-generated holograms, *Proc. IEEE* (*Lett.*) **55** (1), 109–111 (1967).
21. W. E. Kock, Holograms of x-ray information, *Proc. IEEE* Nov., 1458–1459 (1972).
22. G. S. Johnson, *Atlas of Gallium-67 Scintigraphy*, Plenum Press, New York, 1973.
23. H. H. Barrett, Fresnel zone plate imaging in nuclear medicine, *J. Nucl. Med.* **13**, 382 (1972).
24. L. Mertz and N. O. Young, Fresnel transformation of images, *Proc. Int. Conf. Opt. Instrum. London* 305 (1961).
25. W. L. Rogers, K. S. Han, L. W. Jones, and W. H. Beierwaltes, Use of incoherent holography for gamma-ray imaging, *J. Nucl. Med.* **13**, 464 (1972).

26. *Bus. Week* **July 15**, 50–52 (1972).
27. *Microwaves* **9**, Aug., 76 (1970).
28. *Optical Spectra* **8**, Jan., 36 (1974).
29. *Lasersphere* **1**, Mar. 8, 1–2 (1971).
30. *Optical Spectra* **8**, Jan., 36 (1974).
31. *Electronics* **44**, July 5, 24 (1971).
32. *Ind. Res.* **14** (6), 37 (1972).
33. T. E. Gordon, Jr. and D. L. Smith, A laser in the dental lab, *Laser Focus* **6**, June, 37–39 (1970).
34. Laser Focus **9**, Nov., 36 (1973).
35. Laser Focus **8**, Nov. 23 (1972).
36. Holosphere **3**, Jan., 3 (1974).
37. Electronics **40**, May 15, 48 (1967).
38. L. Goldman, R. J. Rockwell, Jr., G. Nath, and G. Schidler, Recent advances in laser surgery using Nd:YAG, *4th Electro-Opt. Conf. Sept. 1972.*
39. *Lasersphere* **2**, Oct. 15, 12 (1972).
40. L. Bush, Laser beam surgery, *Ann Arbor News*, **Dec. 20**, (1970).

14

COMPUTING APPLICATIONS

One important field of laser application involves their use in optical processing, an area now often referred to as optical computing. In this field use is often made of the Fourier transform property of a lens and the associated filtering procedure called *spatial* filtering. To introduce these and related concepts dealing with the electro-optical processing of information, we will review several applications of Fourier analysis, Fourier series, and the Fourier transform.

14.1. Fourier Series

The Fourier series is used to describe periodic functions, presenting them as infinite sums of oscillations at frequencies that are harmonics (discrete spectrum) of the fundamental. The Fourier transform can express an arbitrary aperiodic function as an infinite integral over a continuous range of frequencies. Often used in the treatment of single-pulse phenomena by the electrical engineer, the Fourier transform now finds application in optics, acoustics, and aperiodic effects in electrical circuits.

The frequency content (i.e., the spectrum) of a constant-amplitude pulse of finite duration can be defined through the use of a $(\sin x)/x$ function. Thus, the two top sketches in Figure 14.1 show how the constant-amplitude pulse E (shown at the left in the time domain) can be represented in the frequency domain (right) as its Fourier transform [the $(\sin x)/x$ function]. The lower two

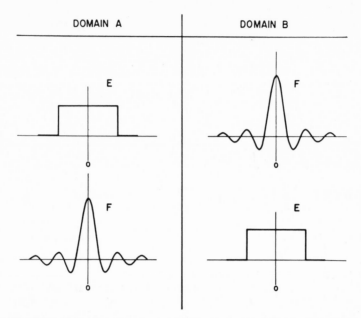

FIGURE 14.1. The Fourier transform and retransform of a pulse.

sketches in the figure portray the fact that the transform of a function is the original function itself.

When uniform-amplitude coherent light waves pass through a slit, as in Figures 1.10 and 1.12, the distant (Fraunhofer)-field pattern is also defined by the $(\sin x)/x$ function (Figure 1.12). Similarly, when a radio antenna, having a uniform intensity of excitation across its aperture, transmits a single-frequency signal, the distant-field pattern is again expressed by the $(\sin x)/x$ function.

The concept of Fourier analysis was first described by J. B. J. Fourier (1768–1830) in his book *La Théorie du Chaleur*. The Fourier analysis of a given periodic function represents it as a sum of a number, usually infinite, of simple harmonic components. We saw such an analysis of two periodic sounds in Figures 2.1 and 2.2. The sound in Figure 2.1 is that of a pure tone having no harmonics. The sound of Figure 2.2 comprises a fundamental at 200 cycles per second (Hz) and five additional harmonically related overtones. Fourier showed that any function $F(t)$ which repeats itself periodically f times per second, such that $F(t + 1/f) = F(t)$, can be represented as a summation of a harmonic series of pendular vibrations

$$F(t) = \sum_0^\infty b_n \cos n\omega t + \sum_1^\infty c_n \sin n\omega t \qquad (14.1)$$

where

$$\omega = 2\pi f$$

$$b_0 = \omega/2\pi \int_0^{2\pi/\omega} F(t)\, dt$$

$$b_n = \omega/\pi \int_0^{2\pi/\omega} F(t) \cos n\omega t\, dt$$

and

$$c_n = \omega/\pi \int_0^{2\pi/\omega} F(t) \sin n\omega t\, dt$$

The same expression employing exponentials becomes

$$F(t) = \sum_{-\infty}^{+\infty} a_n \exp(ni\omega t)$$

where

$$a_n = \omega/2\pi \int_0^{2\pi/\omega} f(t) \exp(-ni\omega t) \tag{14.2}$$

In Figure 2.2, the amplitudes are not indicated, but the frequencies f would be as shown and six terms would thus completely define this periodic sound.

When a singly resonant electrical circuit is repeatedly excited by a short pulse, the resonance frequency is generated, dying away as shown in the top left of Figure 14.2, until the next pulse again excites the circuit. The frequency and the damping of the resonant circuit can be obtained from the outline of the harmonics in the Fourier series pattern of such periodic waves. This is evident in the harmonic analysis in Figure 14.2. It has also been shown [1] that a mechanical resonance in a musical instrument can cause the Fourier spectrum to display the properties of such a resonance (called a formant) at *various* repetition rates of the sound pulses generated in the mouthpiece of the instrument. Figure 14.3 shows this effect. [2]

The analyses portrayed in Figure 14.3 were obtained by scanning the frequency range of interest with a very narrow filter; the filter response rose as the filter passed each harmonic. A technique for portraying the Fourier analysis of *changing* signals was developed by Potter [3] for use in analyzing and portraying the varying sounds of speech. Figure 14.4 portrays such a varying vocal sound, with the *amplitudes* of the harmonics here being indicated by the darkness of their (horizontal) patterns and their frequency plotted vertically; time is shown in the horizontal direction. Figure 14.5 shows a similar analysis of the spectrum of a varying-speed diesel engine. Here the frequency scale is smaller, to permit the much lower fundamental of the engine, and its harmonics, to be portrayed.

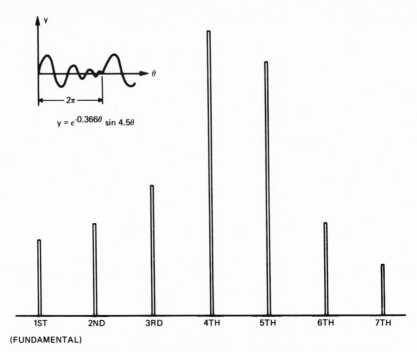

FIGURE 14.2. An example of a Fourier analysis of a repeated signal.

14.2. Discrete Fourier Analysis

In the Fourier equations 14.1 and 14.2, the function to be analyzed must be analytically expressible to permit the series coefficients to be computed. When the repeated wave cannot be expressed analytically, another process is available. In this procedure, samples are taken at discrete intervals along the course of the wave period, and the wave amplitudes at those points (Figure 14.6) are used to determine (approximately) the strength of the harmonics. This process, called the *discrete* Fourier analysis, requires very many samples to be taken if the wave form contains steep portions. This is because such portions correspond to a very high harmonic content and the number of the required samples is proportional to the number of harmonics that must be defined.

This procedure for determining the harmonic content involves large numbers of multiplications of the sampled values, with the values of the sine and cosine functions corresponding to the position of the sample (in degrees) in the wave period, and also numerous additions and subtractions.

BASSOON

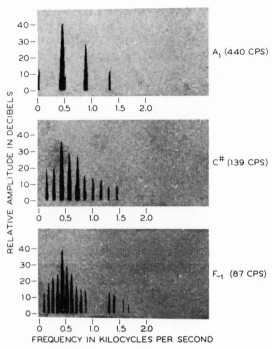

FIGURE 14.3. The musical instrument resonance at about 500 cycles per second (Hz) modifies the harmonic structure at all fundamental frequencies.

14.3. Discrete Analysis Tables

Near the turn of the century,[4] the German mathematician G. Runge described a method for substantially reducing the great amount of labor involved in the direct determination of the coefficients when many samples (ordinates) are necessary.

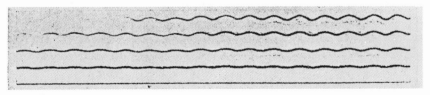

FIGURE 14.4. Four harmonics of a soprano's voice showing the periodic frequency fluctuation (vibrato) (from Reference 2).

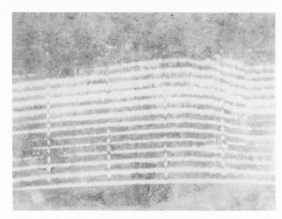

FIGURE 14.5. The harmonics of a diesel engine vary in frequency as the engine speed changes.

Runge's method involves arranging the various operations in tabular form, and making use of the fact that many of the sine and cosine functions are the same (e.g., cos 240° = cos 120°, cos 300° = cos 60°, etc.). Noting that the normal process involved large numbers of *repetitive* multiplications, Runge worked out tabular forms or schedules, in which many additions and subtractions were performed *first*, with these being followed by a much smaller number of multiplications with the trigonometric values. An example of such a table [5] is shown in Figure 14.7. Lipka [6] has noted that in determining the amplitude of *six* harmonics by the use of such a table, the time and labor spent is less than that required for determining *one* harmonic by

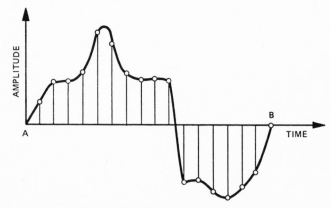

FIGURE 14.6. In a discrete Fourier analysis, the wave period is sampled.

(1) $f(x) = b_0 + a_1 \sin x + a_2 \sin 2x + a_3 \sin 3x + \cdots + b_1 \cos x + b_2 \cos 2x + \cdots$

(2) $f(x) = b_0 + c_1 \sin(x + \alpha_1) + c_2 \sin(2x + \alpha_2) + c_3 \sin(3x + \alpha_3) + \cdots$

Checks: $Y_0 = b_0 + \sum b_n$, $0.5(Y_1 - Y_{23}) = 0.257(a_1 - a_{11}) + 0.5(a_2 + a_{10}) + 0.707(a_3 + a_9) + 0.866(a_4 + a_8)$
$+ 0.966(a_5 + a_7) + a_6$

	Y_0	Y_1	Y_2	Y_3	Y_4	Y_5	Y_6	
	Y_{12}	Y_{11}	Y_{10}	Y_9	Y_8	Y_7	Y_{18}	
Sum	s_0	s_1	s_2	s_3	s_4	s_5	s_{11}	$\mu_1 = T_6 + T_{10} =$
Diff.	d_0	d_1	d_2	d_3	d_4	d_5	d_{11}	$\mu_2 = T_7 + T_9 \times 2$
		Y_{13}	Y_{14}	Y_{15}	Y_{16}	Y_{17}		$\mu_3 = T_2 - T_4 =$
		Y_{23}	Y_{22}	Y_{21}	Y_{20}	Y_{19}		$\mu_4 = T_1 - T_5 =$
Sum		s_6	s_7	s_8	s_9	s_{10}		$\mu_5 = s_0 - s_{11} =$
Diff.		d_6	d_7	d_8	d_9	d_{10}		$V_1 = R_1 + R_3 - R_5 =$
		s_1	s_2	s_3	s_4	s_5		$V_2 = R_2 - d_{11} =$
		s_6	s_7	s_8	s_9	s_{10}		$V_3 = R_6 - R_8 - R_{10} =$
Sum		T_1	T_2	T_3	T_4	T_5		$V_4 = d_0 - R_9 =$
Diff.		R_1	R_2	R_3	R_4	R_5		$W_1 = T_6 - T_{10} =$
		d_1	d_2	d_3	d_4	d_5		$W_2 = T_7 - T_9 =$
		d_6	d_7	d_8	d_9	d_{10}		$W_3 = T_2 + T_4 =$
Sum		T_6	T_7	T_8	T_9	T_{10}		$W_4 = T_1 + T_5 =$
Diff.		R_6	R_7	R_8	R_9	R_{10}		$W_5 = s_0 + s_{11} =$

FIGURE 14.7. A discrete harmonic analysis table or schedule.

the direct process. Runge, thus, in 1903, provided the first speed-up of the Fourier process, a *faster* Fourier analysis.

14.4. Fourier Integrals

In Figure 2.2 the spectrum of a sound which is noiselike and therefore not periodic is portrayed. For this sound, the Fourier series analysis technique cannot be used to specify its acoustic content. However, an extension of the Fourier series procedure which involves Fourier *integrals* can be used to represent such nonperiodic sounds (or periodic sounds having a very short duration). Thus, a nonperiodic function can be considered as a periodic one having an infinitely long period. Accordingly, instead of integrating over one period (from 0 to $2\pi/\omega$), as is done in obtaining the Fourier series, the limits of integration are extended from $-\infty$ to $+\infty$, and the period $2\pi/\omega$ is extended to ∞, or $\omega/2\pi \to 0$. In this way, the nonperiodic function $F(t)$ can be expressed as the sum of a continuous spectrum of periodic functions spaced infinitely close together in frequency.

14.5. Fourier Transforms

When the quantities F and f are related by the formula

$$F(t) = \int_{+\infty}^{-\infty} f(z) \exp(2\pi i z t) \, dz \qquad (14.3)$$

then F is said to be the Fourier transform of f. The close relationship to the Fourier integral is evident; in fact, today the term Fourier transform (particularly the term fast Fourier transform) is often applied to both the Fourier analysis and the Fourier integral processes.

14.6. Fast Fourier Transforms (FFT)

The rapid growth of digital computers in the last two decades has led to ways being formulated for accomplishing digitally many mathematical operations. It was recognized that the performing of Fourier analyses and Fourier integral processes by a computer would be very valuable, but because of the digital (nonanalog) nature of these computers, it was the *discrete* version of the operation (Figure 14.6) that had to be chosen for such operations. Several successful methods resulted, with the most widely used programing for this operation having been devised by J. W. Cooley and J. W.

Tukey in 1965.[7] It achieves, as did the Runge tables, a significant saving in time by the prior combination of addition and subtraction operations. When used for a discrete Fourier wave analysis involving eight samplings of the wave period, the normally required 64 complex multiply and add operations are reduced to 12.[8] This major programing step is now referred to as the Cooley–Tukey algorithm. The scientifically oriented computer programing languages are called algorithmic, with the language ALGOL, which is based on the notation of elementary algebra, now being the almost universally used language (*alg*ebra *o*riented *l*anguage). We saw in Figure 10.35 a sample of an FFT frequency–time plot of an underwater acoustic interference pattern.

14.7. Distant-Field Patterns

We noted at the beginning of this chapter that the Fourier transform, in addition to its use in signal analysis, can also provide information about the distant-field pattern of an optical slit (Figure 1.10) or a radio transmitting or receiving antenna. Thus, whereas the $(\sin x)/x$ pattern of a uniformly illuminated slit (Figure 1.12) has secondary maxima that are rather high (the first minor lobes are only 13 dB lower than the main lobe), the pattern from an aperture for which the amplitude distribution is *tapered* possesses much lower side lobes. The Fourier transform permits a determination of the extent of this reduction if the taper function is analytically expressible. Thus, if the amplitude of illumination is made to follow a half-sine-wave taper, the function and the integral are readily calculable. Figure 14.8 shows the calculated radiation pattern formed by two spaced radiators; the scanned pattern is shown in Figure 1.15.[9] The similarity between the two patterns is quite apparent.

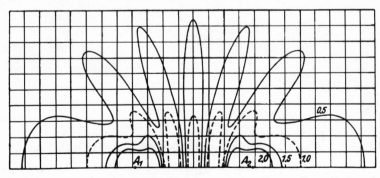

FIGURE 14.8. Calculated pattern of two spaced radiators (from Reference 9).

14.8. Near-Field Patterns

We saw that the patterns of synthetic-aperture systems (and those of holograms) fall in the *near-field* category, as do also those of transducers often used in ultrasonic medical diagnostics. Figure 14.9 shows a Schlieren pattern of the near field of an ultrasonic transducer;[10] the similarity to the pattern of Figure 1.14 is quite noticeable. Figure 14.10 is a repeat of Figure 1.14, with the reference wave having been added. It shows why the pattern *is* a near-field one; the straight, parallel wave fronts show that the waves are still moving straight ahead, rather than curving outward as they would when the distance to the far field is reached, i.e., as the waves from the horn in Figure 10.3 do. The phase pattern was made strong in this record, but the near-field variation can still be seen.

Near-field patterns can also be calculated; Figure 14.11 shows an experimentally recorded near-field pattern superimposed on the calculated pattern.[11] This would correspond to one of the intermediate points between the second and third lower curves in the vertical set of patterns shown at the right of Figure 1.13.

FIGURE 14.9. Photograph of a Schlieren pattern of an acoustic radiator positioned at the left of the figure.

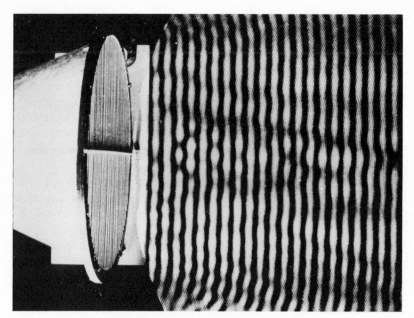

FIGURE 14.10. The phase pattern of Figure 1.14.

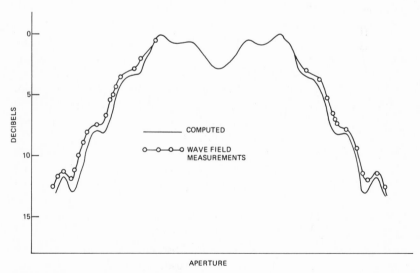

FIGURE 14.11. A computed and experimentally measured near-field pattern.

14.9. Space–Frequency Equivalence

The Fourier transform operation is also useful when correlation methods are used for detecting signals. In one procedure, two *spaced* receivers are used, with the desired result being the detection of a target radiating a broad-band noise signal. The outputs of the receivers are multiplied, then rectified, and the rectified signal smoothed. When a noise source is centrally located along a line perpendicular to the line joining the two receivers, both received signals are identical, and the multiplication and smoothing process results in a strong output signal.

It has been shown[12–15] that the target signal need not be a true noise source, but could be rather a cluster of single frequencies, with their spacing being determined by a method exactly analogous to that of choosing the spacing of array elements in a single-frequency system. Figure 14.12 shows this space–frequency concept with the third and fourth cases indicating this effect. Figure 14.13 depicts an experimental demonstration of this equivalence.[12]

14.10. Two-Dimensional Fourier Transforms

Up to this point, we have discussed one-dimensional problems, including periodic or single-pulsed signals, and the one-dimensional radiator aperture patterns and their transforms. However, most ultrasonic sonar transducer arrays, such as the one shown in Figure 14.14, and many microwave array radars, such as the very large one shown in Figure 14.15, have two-dimensional apertures. Such systems have radiation patterns comparable to the pattern of Figure 14.16, although it would be unusual for them to have the exact symmetry shown in the pattern of that figure. It is in such two-dimensional situations where optical techniques, using lenses for generating two dimensional Fourier transforms of a function, are found particularly useful. As was stated in a paper by IBM authors[16] on kinoforms (a subject we discussed in Chapter 10), "The digital computer provides a fast, accurate way of operating on serial data, but difficulties arise in the application of the digital computer to two-dimensional image processing problems."

The ability of coherent light to transform a two-dimensional electric field pattern existing in the pupil plane of a perfectly corrected (ideal) lens system into its spatial Fourier transform in the image plane of the lens[17,18] has brought the laser into numerous applications in the computing field. The fact that a laser can accomplish this two-dimensional transform almost instantaneously has led to several interesting titles of papers, such as "Com-

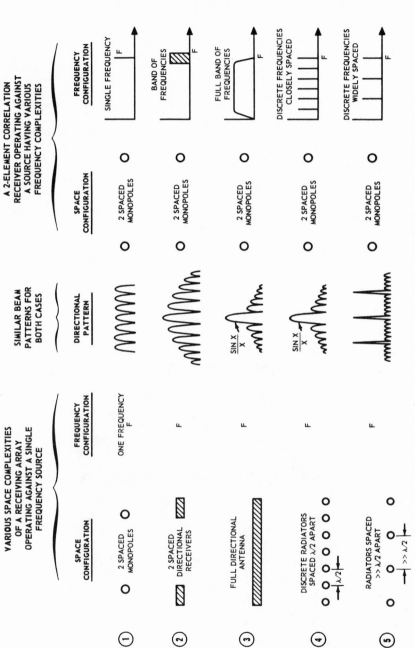

FIGURE 14.12. Space–frequency equivalence. Examples 3 and 4 are of particular interest.

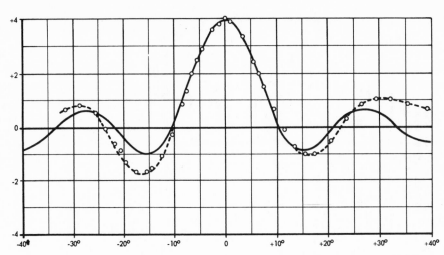

FIGURE 14.13. Experimental verification of space–frequency equivalence (dashed line experimental).

FIGURE 14.14. A multielement, two-dimensional sonar array.

FIGURE 14.15. A multielement array radar. The two objects in the center are workmen (courtesy of Bendix Corp.).

puting at the Speed of Light."[19] Similarly, S. P. McGrew has noted that when two data planes, each having one billion data points, are positioned adjacent to one another in the Fourier transform plane, each of the adjacent two data points is multiplied, resulting in a billion multiplications at the speed of light, i.e., 10^{18} multiplications per second.[20]

14.11. The History of Optical Processing

Of all the techniques now in use for the handling of information, optical processing is the oldest from the standpoint of coverage in the literature and the newest from the standpoint of practical use. It has been a century since Ernst Abbe[21] published a classic paper demonstrating how information on a picture can be filtered by forming a diffraction pattern and by blocking appropriate portions of the light in the pattern. He showed that this is possible because the diffraction pattern is the two-dimensional Fourier transform of the light distribution in the picture. As shown in Figure 14.17, the input pattern existing in the object plane is transformed into the spatial frequency domain (the transform plane). At this point various operations can be performed. Thus, one can sample the Fourier plane and thereby ascertain the Fourier coefficients. Another application involves the insertion of a spatial filter in the transform plane with the filtered pattern then being retransformed

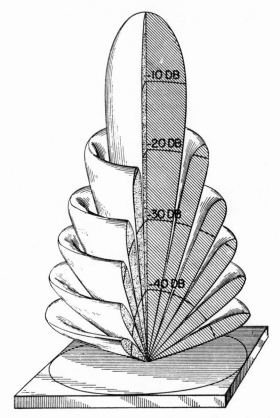

FIGURE 14.16. Idealized pattern of a two-dimensional array.

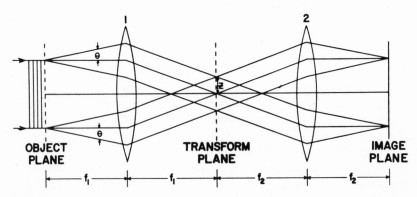

FIGURE 14.17. A lens from the Fourier transform of a two-dimensional signal located in the object plane. A second lens retransforms this transform back into the original signal in the image plane.

into the image plane. In this way, one can implement a pattern recognition system, since the proper filter placed in the transform plane will permit only the desired patterns present in the object plane picture to be retransformed back into their original patterns.

Over the past years, a number of authors such as A. Marechal and P. Croce,[22] L. J. Cutrona,[23] and E. L. O'Neill[24] have shown how Abbe's concept could be related to modern information-processing principles. But it was not until the gas laser was introduced in the early 1960's that it became feasible to use optical methods for processing data on a routine production basis. In contrast to electrical analog and digital processing, the extent to which optical data-handling systems are being employed is still quite limited. As we noted, the primary advantages of optical techniques over the other two are best realized when the information to be processed has two degrees of freedom. A picture, for example, has two independent variables (the x and y position coordinates) and one dependent variable (the density at the position specified). Optical systems, correspondingly, have two degrees of freedom; thus, for handling pictorial data, they are intrinsically superior to electrical or electronic systems, which have only time as an independent variable.

14.12. Optical Filtering

If the information to be processed is transferred to film and if a beam of coherent light, such as from a laser, is passed through it (Figure 14.17), a lens situated a focal length beyond the film transparency will convert the information into a diffraction pattern that, as Abbe showed, has all the properties of the Fourier transform of the input information. Another lens, which is beyond the transform plane a distance equal to its own focal length, will convert the diffraction pattern into a second Fourier transform. But, as the transform of a transform of a function is the original function itself, the final image, or reconstruction, is the same as the original input reduced or enlarged by a magnification factor. If one blocks the light focused in those portions of the transform plane corresponding to undesired spatial frequencies or directions of orientation, the final image will then show all the information on the original input except for that which has been removed in the transform plane. This phenomenon is the basis for optical filtering.

The transform itself presents the input information in a form that is exceptionally convenient for frequency analysis; it is a simple matter to photograph it for this purpose, because photographic film responds to the square of the light intensity and because the phase term of the transform cannot be preserved on the photograph without the introduction of a

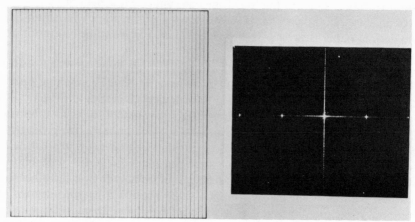

FIGURE 14.18. The Fourier transform of a vertical line array (left) is a series of dots of light (right).

reference beam like that in holography, the transform becomes the two-dimensional power spectrum of the information in the output.[25]

At the left of Figure 14.18 is a series of parallel, equally spaced, vertical lines. On the right is the diffraction pattern, or transform, as generated by the set of lines at the left. The set of lines can be looked upon as an optical grating, so that in the transform section, we see (1) a zero-order diffracted component (a component having a spatial frequency of zero), (2) a pair of first-order diffracted components (the two bright dots nearest to the central spot), and (3) a pair of second-order components (the other two bright dots).

Returning to Figure 14.8, we recall that plane waves are arriving from the left, and we note that a horizontal line grating is sketched in the object plane. The two first-order diffractions are shown, with the diffraction angle, upward and downward, indicated as θ. The two upward-deviated rays at the far left are made to converge (by lens 1) to a *point* in the transform plane which is at a distance Z above the central axis. *All* portions of that upwardly diffracted plane wave converge to this same point. The other two sets of plane waves are similarly made to converge to dots in the focal plane of the lens (the transform plane); these three dots correspond to items (1) and (2) of the previous paragraph. If we were now to block the transmission of the light passing through the four diffraction dot positions of Figure 14.18 (through the use of opaque material), the retransformed signal at the right of Figure 14.17 would be free of the vertical lines which exist in the object plane and which generated the focused spots. Other information which might have been present in the right portion of Figure 14.18 would not have been affected by the opaque-dot spatial filter; only the vertical lines would have been filtered out.

14.13. Seismic Applications

The optical filtering procedure was applied quite early for the improvement of seismic records.[26] In such records, various forms of noise or other artifacts often obscure the low-frequency sound-wave reflection pattern which could otherwise indicate the presence of underground oil or gas formations. Optical filtering systems for improving the quality of the records (Figure 14.19) are provided with filters which can pass or reject any desired range of spatial frequencies (or any desired range of inclinations) present in the original picture. Thus, if low spatial frequencies are to be eliminated at all orientations, one masks the area around the central axis with an opaque disk; its cutoff frequency is determined by the radius of the disk. High frequencies can be removed by placing an iris diaphragm in the transform plane. The term "inclinations" refers to seismic echoes reflected from tilted layers or tilted rock formations existing in the earth where the seismic exploration is being conducted. Such a condition is stimulated in Figure 14.20, where a set of tilted plane reflections is shown (left) as a series of heavy bars. In this case, the transform plane (right) exhibits not only focused spots positioned along the vertical and horizontal axes (corresponding to the vertical and horizontal lines in the left figure), but also a set positioned along a tilted line (perpendicular to the heavy lines in the left figure). To remove *these* heavy, tilted lines, an opaque, *tilted*, sectorlike filter (referred to as a directional rejection filter) is used.

FIGURE 14.19. The optical processor called the Laserscan, used in seismic exploration.

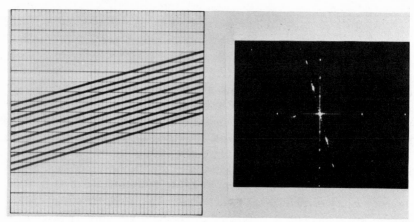

FIGURE 14.20. Three sets of parallel lines (left) generate three sets of light spots (right).

14.14. Complex Optical Filters

Up to this point we have not considered the vectorial nature of (electromagnetic) light waves, yet to describe *exactly* the diffraction of light waves, the polarization of the wave must be known. Microwave engineers are cognizant of the fact that the diffracted wave pattern at a knife edge (Figure 1.7) for horizontally polarized waves is *not* the same as the pattern for vertically polarized waves, and that, because of this, other diffracting devices such as two-dimensional zone plates, generate, for polarized waves, patterns in the horizontal plane which are different from the vertical plane patterns. Stroke has discussed this effect which exists in closely spaced optical gratings.[27,28] The wide-angle zone plates of holography often fall in a category which is similar to that of closely spaced gratings.

However, the rigid vectorial treatment is quite complicated, and the simpler scalar theory can handle quite satisfactorily the cases we have discussed. Also, in the discussion thus far of optical filters, we have considered only their *amplitude* characteristics, ignoring their *phase* properties. Many electronic filters, however, provide for their desired function both the needed amplitude and phase characteristics. Thus, a *matched* filter has a transmittance which is proportional to the complex conjugate of the spectrum of the signal to be filtered. To accomplish this in an optical filter, the *phase* of the signal spectrum must be preserved and operated on.

A method for doing this was described quite early by H. Vander Lugt.[29] Because a Gabor hologram can be considered as a way of recording the phase information present in the object wave, the making of a *hologram* of the (spatial) transform spectrum permits the phase information to be made available. Thus, if in Figure 14.17, a plane, off-axis, reference wave is directed at the transform plane, and a photographic plate positioned in that plane, a hologram would be made of the Fourier transform of the two-dimensional information existing in the object plane. If this transform has phase information (and any asymmetry in the object plane signal will cause the transform to be a complex signal), the hologram process records this phase information.

If, after developing and fixing, the "hologram" is again positioned in the transform plane of Figure 14.17, and a *point* source of coherent light is placed at the center of the object plane, the hologram will be illuminated with plane waves, and the impulse response of the filter will be displayed as an off-axis image generated in the image plane. The conjugate of the impulse response will also appear as an off-axis, *inverted* image. J. Tsujiuchi demonstrated in 1963 that the required complex filters could be realized by a combination of two; the amplitude filter, realized by photography, and the phase filter, realized by vacuum evaporation.[30] Also, both Leith[31] and Stroke[32] noted that the complex filter could be made up of two planar components (phase and amplitude).

14.15. Complex-Filter Applications

In x-ray astronomy, because of the inability to provide focusing of the x rays (through lenses or concave mirrors), single, pinhole cameras were generally used. In 1968, R. H. Dicke proposed the use of an aperture having not one, but a large number of (randomly disposed) pinholes.[33] The energy transmission through *this* "camera lens" would obviously be higher than for the single, pinhole aperture, and Dicke noted that the required single-image synthesis from this multiple-image picture, could be accomplished using coherent light in an optical Fourier-transforming arrangement, provided that an "appropriate phase—and amplitude correcting plate"[33] could somehow be devised.[34] Using the Fourier-transform-hologram technique for making the desired correcting plate, Stroke showed that the x-ray signals from all of the many pinholes could indeed be recovered and *superimposed* so as to form a high-gain, sharp picture.[34]

Along the same lines, Stroke has utilized Fourier-transform filters in many ways, with good success in correcting blurred photos, i.e., in "deblurring" them. As shown in Figure 14.21, the Fourier-transform hologram is first made of the blurred photograph (top left) by using a very sharp *point source* for the reference illumination. If the same point source were again used

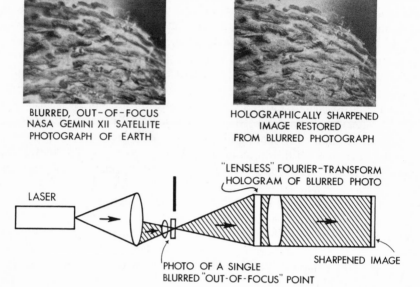

BLURRED, OUT-OF-FOCUS
NASA GEMINI XII SATELLITE
PHOTOGRAPH OF EARTH

HOLOGRAPHICALLY SHARPENED
IMAGE RESTORED
FROM BLURRED PHOTOGRAPH

FIGURE 14.21. Fourier transform filters can "deblur" blurred photos (from Reference 34).

to illuminate the hologram, the same blurred photograph would be reconstructed. Instead, as shown in the lower part of the figure, a point source of light which is itself blurred (out of focus) is used in retransforming the Fourier-transform hologram into a sharpened image.[35] When the blurring is caused by *motion*, the deblurring process uses a filter matching the blur pattern of a point which has become blurred into a short line. Figure 14.22 shows the improvement in this case.[36]

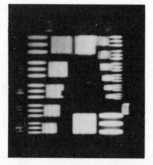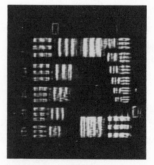

FIGURE 14.22. A photograph blurred by motion (left) and the holographically deblurred photograph (right) obtained from it (from Reference 36).

Complex optical filters have also been used in the pulse-compression process, which we discussed earlier, known as chirp. The uncompressed pulse, a long, linear, FM signal, is positioned in the object plane where, with laser illumination, it can be made to exhibit, in the transform plane, its Fourier transform. This spatial-transform signal can be converted into a plane wave in that transform plane by placing a planoconvex lens there. The resulting *plane* waves are brought to a point focus by the second lens (as in Figure 14.17), resulting in the long FM chirp pulse being significantly compressed in length.[37] Leith has discussed similar optical-processing techniques for pulse compression.[38]

The subject of holographic deblurring and complex optical filtering has been treated at length by Stroke,[39] this paper also includes a rather extensive reference list (51 items).

14.16. Optical Processing in Radar

We noted in Chapter 9 that the coherent form of radar called synthetic-aperture radar utilized optical-processing techniques for placing the received data in usable form. This retrieval of the radar information about the terrain flown over by the aircraft employs the hologram process, and, as the classic Cutrona *et al.* paper[40] on synthetic aperture stated, the radar utilizes "the spatial Fourier transform relation that exists between the complex light-amplitude distributions at the front and back focal planes of a lens." As was pointed out in Chapter 9 the hologram nature of the process was not noted in this paper. It was stated that "the required processing can be viewed as a multichannel weighted summation, a matched filtering operation or a cross-correlation operation; the cross-correlation viewpoint is developed as is a corresponding optical processor implementation". The similarity of the records to one-dimensional zone plates was noted, but this was tied to the optical processor rather than to any hologram similarity. Thus, it was stated that, "A Fresnel lens (zone plate) interpretation of the recorded radar signal leads to a discussion of the strong physical analogy between a coherent radar system and a coherent optical system." Also, no indication was given of the extremely high degree of coherence required in holography. At the 1967 U.S.–Japan seminar on Holography sponsored by the National Science Foundation, this author stressed this hologram analogy, and the high coherency required. He published a note on this early in 1968 and a more complete paper later in that year.

The optical-processing aspects of this form of radar led to a rapid extension of optical-computing possibilities, and Stroke refers to this in the section "Historical Background" of his 1972 optical-computing paper.[39]

14.17. Parallel Optical Processing: The Accepting Function

It was not long after the early discussion of the holographic computing aspects of the synthetic-aperture technique[41] that the really quite surprising and unexpected computing properties of this holographic technique began to be recognized. We will review here some of these properties as described earlier.[42]

As we have noted, the great equivalent length of the synthetic (linear) aperture permits many extremely sharp beams to be generated. In most radars which have such large apertures, it is an extremely complicated procedure to cause all of these beams to be formed simultaneously. Instead, *one* sharp beam is usually formed and this beam is scanned (serially) over the many possible directions in which it can be pointed. In the synthetic system, all beams are automatically formed (and information is accepted by all) simultaneously.

Any optical or radar aperture whose dimensions extend over many thousands of wavelengths has a near field which extends outward to very great distances. For such an aperture to have maximum effectiveness in delineating the large number of objects in its near field, the elements must be phased so as to exhibit a *curved*-wave-front property. This curvature is very great for near objects, is less so for midrange objects, and becomes flat for far-field objects. For the standard form of radar, such a phase-adjustment procedure, differing for each element of range, would be a very complicated one, and if possible at all, it would surely be done in a serial manner. In the synthetic system, such curvature is automatically introduced through the zone-plate action. Information is thus accepted from all range points in a highly efficient manner and in parallel.

In certain radar applications it is desirable to have the metric (not the angular) resolution constant for all ranges. For this to occur, a much larger aperture must be utilized for more distant points (thereby providing a much higher *angular* resolution) than for near points. In an ordinary radar, the procedure to accomplish this is again a very complicated one. In synthetic-aperture systems, this change of aperture length for the various ranges also occurs automatically because of the finite (small) aperture size of the actual antenna employed. *Its* constant angular resolution *versus* range (usually tens of degrees) causes *its* metric resolution to be less at great ranges. This provides *longer* zone plates for the more distant objects and *shorter* ones for the nearby objects. The result is a metric resolution which is independent of range. Parallel acceptance of information is automatically provided, thereby achieving the desired result.

Finally, in the accepting function, it is of interest to note that the beam width of a synthetic aperture is twice as sharp as that of a standard radar of equal aperture.

14.18. Parallel Optical Processing: The Processing and Display Function

The processing of the received data in a synthetic-aperture system can be considered to be, as in a hologram, the photographic recording procedure, which results in myriads of superimposed zone plates and in the developing, in parallel, of all of the zone plates in that photographic record.

As in the holographic reconstruction of a scene, the display function of a synthetic-aperture system is accomplished by the illumination of the photographic record with laser light, causing the parallel displaying of myriads of objects located at many different points in range.[43]

14.19. Real-Time Optical Computing

As we noted in Chapter 13, in holography and synthetic-aperture systems there is a time delay because of the requirement to develop and reconstruct the hologram. We there mentioned briefly the possibility of using the Goetz tube (Figure 9.6), in which the microwave (or ultrasonic) holograms can be recorded and quickly reconstructed in real time, noting that in early models the real-time capabilities were of the order of 16 frames per second.

As was shown in Figure 9.25, all of the one-dimensional zone plates of a synthetic-aperture record can also cause laser light to be brought to a focus on the (tilted) plane of reconstructed real images. The focal *region* of each such zone plate is actually a circle, but because of the directionality of the plane-wave, laser, illuminating light, the point of that circle where it intersects the tilted plane is very bright. These focal points correspond exactly to the range and azimuth of the reflecting points which generated the zone plates.

It was recently suggested[44] that the information shown on the hologram plate of Figure 9.25 be recorded (electronically) on the crystal of the real-time tube of Figure 9.6, thereby permitting that (changing) hologram information to be repeatedly recorded and reconstructed at sufficiently rapid repetition rates (e.g., 16–30 times per second) to provide a real-time visual readout. The changing view would appear on a tilted (possibly ground-glass) plane similar to the plane of Figure 9.25. In such radar, successive views of the terrain would be available to the aircraft pilot in real time, even during heavy fog or cloud cover, and by aiming two radars, one on each side of the aircraft, toward the *forward* portion of the normal side-looking view, the all-weather views could prove useful in commercial-aviation applications (even though the pictures would obviously not show the fine detail of Figure 9.29). Even in the normal use of present synthetic-aperture radars where the real-time feature is not essential, a monitor based on this concept could be useful. As mentioned in Chapter 13, such a real-time device would also be useful in ultrasonic synthetic-aperture medical diagnostics. It was suggested[44] that the tube (the optical computer) could be operated in a digital mode, with signals restricted to the binary values of 0 and 1. The system could then employ the digital holographic logic successfully demonstrated by K. Preston.[45]

Other real-time tubes for holographic computing (optical processing in real time) include one under development by the Itek Corp., called the "PROM" (an acronym for Pockels *readout optical modulator*).[46] It will be recalled that in Chapter 9, in describing the operation of the Goetz tube, we mentioned the "electro-optic" effect; this is also referred to as the Pockels effect. The developers feel that the tube will be useful through its ability to provide the *parallel* data-handling capability of optical computing. Another real-time tube, under development by Phillips Research in Europe, is called the TITUS.[47] Also, D. Casasent has had good success with the EALM (the *electronically addressed light modulator*).[48]

14.20. Computer-Generated Holograms

A considerable interest has arisen in exploring the possibilities of applying the holographic principles of three-dimensional image storage and reconstruction to digital-computer displays and to other computer applications. With such techniques it is possible to generate holograms capable of displaying three-dimensional images of objects which never existed in reality.

Methods have been described by A. W. Lohmann and B. R. Brown for synthesizing holograms by photographically reducing the intensity distribution obtained in a large-scale drawing, obtained by a computer-guided plotter.[49] Such digital computation of a hologram is in general very complicated, except in simple cases. In a more recent paper, Lohmann describes a technique involving over 16,000 apertures.[50] Figure 14.23 shows an image reconstructed from a hologram made by this process, showing the resolution which is possible, the gray scale, and the ability to portray noise from a simulated diffuser. This and other binary holograms used to reconstruct images were generated in a 128 × 128 array. They were computed with an IBM system 360 Model 50 computer in 4–8 minutes, depending on the object complexity. The 70-cm-square artwork was plotted in three strips to accommodate the 25-cm width of the CalComp 565 incremental plotter. Plotting time was typically 40 minutes. After assembling the strips, a 180× photoreduction onto Kodak 649F film was made in one step using a Leitz Summicron-R 50-mm camera objective. The resulting holograms are 4-mm square, with an average aperture spacing of 0.03 mm.[50]

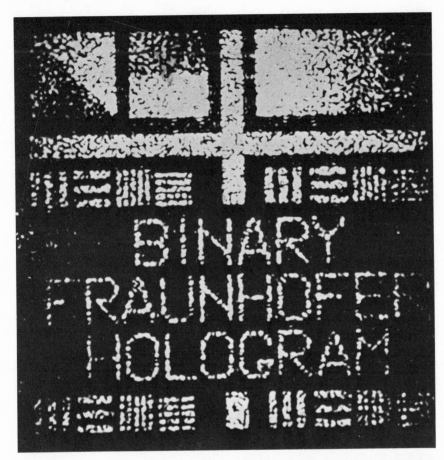

FIGURE 14.23. Reconstruction of a computer-generated hologram (from Reference 5).

As the title of Reference 49 suggests, the possibility of "manufacturing by computer," *complex* spatial filters appears as a very worthwhile goal. Just as in the case of the discrete Fourier analysis (Figure 14.6), the continuous Fourier transform can also be sampled, and with these samples an adequate replica can be constructed. Also, we discussed in Chapter 10 another promising computer-generated hologram, the kinoform.

14.21. Holographic Computer Memories

It was recognized rather early in the development of holography that a hologram has a very large information-storage capability. It can store three-dimensional information, and the information can be stored on the surface or throughout the volume of the hologram. Holograms therefore offer the potential for significant improvements in capacity and access time over existing mass memory technologies such as disks and magnetic tapes; recent advances in optical memory technology make it the leading contender for the next generation of mass storage devices.

The main interest in optical memories stems from the large bit densities that are attainable. The use of lasers offers a significant increase in bit packing density over magnetic recording since the bit size can approach the wavelength of light (0.5μ). In addition to improved density, holographic storage promises a significant reduction in access time over other mass storage systems. The major cost is in the optics, so that the cost of small memories will be quite high on a per-bit basis, but this will drop sharply for large-capacity systems. Accordingly, optical memories are primarily directed at mass storage applications.

Magnetic-disk devices are presently capable of storing approximately 6.4×10^9 bits in a system consisting of eight spindles, each having 20 recording surfaces. Information is stored at a density of 8×10^5 bits/in. (1.2×10^5 bits/cm^2), providing an active recording area of approximately 10^4 in.^{2} (6.5×10^4 cm^2). To extend the capacity of these systems to 10^{12} bits requires either an increase in recording area or an increase in bit packing density by 150 times. Previous experience has shown that mechanical systems for addressing areas of more than approximately 10^4 in.^{2} (6.5×10^4 cm^2) are unreliable and an increase in bit packing density by a factor of 150 is not feasible with the present magnetic technology. Consequently there is strong incentive for introducing a new technology for very large memories.[51]

14.22. Flying-Spot Scanning

Figure 14.24 shows an example of a technique called a holographic flying-spot scanning memory.[52,53] The X–Y deflector directs a laser beam to one of the stored holograms, and the laser light emerging from a hologram illuminates certain ones of the array of photodetectors in accordance with the information that had previously been stored in the holographic memory. This system does not utilize the added information possibilities of the *volume* holograms, but its basic elements can be built with currently available materials and techniques.

The Soviet scientist A. L. Mikaeliane described such a system using an electro-optical X–Y deflector and a photodiode matrix for the information readout. Figure 14.25 shows an image reconstructed from one of his systems holograms.[54]

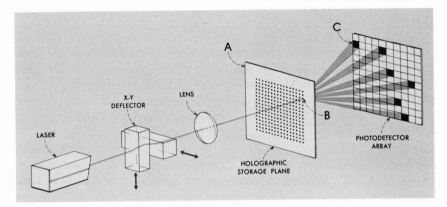

FIGURE 14.24. Sketch of a flying-spot holographic computer memory (courtesy of Bell Laboratories).

14.23. Read-Only Memories

The computer memory system of Figure 14.24 uses a multiplicity of recorded holograms (such as could be placed on a photographic plate), and the operation shown in that figure is a readout of the information in one of the holograms of the memory. Obviously, such an arrangement does not permit certain of the operations which are available in magnetic

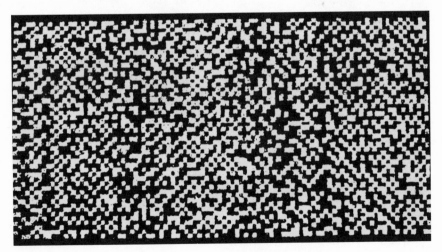

FIGURE 14.25. Reconstructed image of a holographic computer "page" (from Reference 54).

tape memories, such as high-speed erasing and the reinsertion (writing) of new data. Accordingly, the memories of the type shown in Figure 14.24 are referred to as read-only memories (ROM). These are used where large amounts of *permanent* data must be stored, or where infrequent changes are needed. Another application would be one where the data storage is comparable to the microfilming of data; thus the optical storage of 400 pages of information in one square inch has been demonstrated.[55]

For such *permanent* data storage, Precision Instruments supplied a trillion-bit (the 10^{12} figure mentioned above) laser memory to the American Oil Co.,[56] and also one to the University of Illinois for use with the Advanced Research Projects Agency's (ARPA) Illiac IV computer located at the University of Illinois. One of the 450 strips of the Precision Instruments' memory (more recently called UNICON) is shown in Figure 14.26. The lighter portion of the strip has been recorded with approximately one billion bits of data, as much as is contained in a disk file requiring 50 square feet of floor space.[57] A holographic read-only memory called Holoscan, made by Optical Data Systems, employs a 35-mm film.[58] Another ROM called "Megafetch" uses 10- by 10-cm plates, each containing 10^6 microholograms.[59]

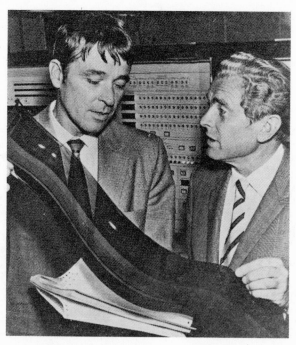

FIGURE 14.26. A flexible "ribbon" which is one portion of a trillion-bit memory (courtesy of Precision Instruments).

14.24. Volume Hologram Memories

Because of the thickness of a photographic emulsion, one element of the emulsion can actually be made to store many different hologram recordings simply by changing the direction of the reference beam for each information unit.[60] In the read operation, to reconstruct the desired *real* image, the conjugate of the reference beam (i.e., an oppositely directed one) is used with the real image then falling on the plane on which the input signal had been during recording. Unfortunately, for the most desirable form of computer memory, i.e., the read–erase–write form, this volume feature is not usable because it is not possible to erase a hologram recorded at one angle without also erasing holograms recorded at other angles.[61] Also the diffraction efficiency of the first hologram to be recorded is reduced as successive holograms are recorded in the same volume element.

14.25. Erasable Memories

Sizable efforts are being applied to materials and techniques which would permit single portions of a hologram memory to be erased and a new signal inserted. In the photographic-film version, the entire memory must be discarded, and another rewritten, when only one tiny portion needs alteration. Materials which were examined quite early for this purpose included photochromic substances such as strontium titanate,[62] thermoplastics,[63] and various ferroelectrics such as lithium niobate.[64]

More recently, these and other substances have been developed to a rather high degree of sophistication. One, called PLZT, is shown in Figure 14.27. It is a transparent lanthanum-modified lead zirconate titanate ceramic, used by both Sandia Labs and Bell Labs in binary and multistate optical memories. Conducting electrodes are deposited so as to enclose or "sandwich" the PLZT ceramic, and when a light pattern falls on the device, depressions are formed in the ceramic, thus causing the image to be recorded. The Murray Hill Bell Laboratories group have named the material FERPIC when used as an erasable storage device for monochromatic images.[65]

Another group of promising materials are the amorphous semiconductors, in which the storage mechanism is based on the amorphous crystalline transition that can be induced in thin films of the material by the heat from the laser illumination. Dr. Neufville of Energy Conversion Devices, the pioneers in amorphous films, has examined their storage possibilities, and IBM was recently awarded a patent on the use of their films of an amorphous chalcogenide.[66] A dye laser is used to write and a krypton laser is used to

FIGURE 14.27. The computer memory substance **PLZT** is almost perfectly transparent.

FIGURE 14.28. Prof. Yu. Nesterikhin of the USSR Academy of Sciences describing an amorphous, erasable, holographic memory unit. (The author has his hand to his chin.)

erase. Soviet scientists have also had good success (Figure 14.28) with chalcogenides.[67]

Arsenic trisulfide (also a chalcogenide) has been found useful by RCA as a holographic memory device.[68] In Figure 14.29 the pencil points to the thin-film hologram at the point where the strong "write" reference beam at the right and the weaker "readout" beam, from the left, cross. This experiment demonstrates that the hologram can be read out as it is being made without any chemical development being required. The holograms are made (exposed) with a low-powered argon (green) laser beam while their images are simultaneously being projected (reconstructed) with an even lower-powered He–Ne (red) laser. The holographer can watch the image develop and turn off the green laser when the image is proper.

More recently, RCA described the use of an improved, thermoplastic photoconductor sandwich for the erasable hologram.[69] An argon laser beam first passes through two acousto-optic crystals that are electronically controlled. The crystals deflect the beam in direct proportion to the frequency of sound waves made to pass through them (Figure 14.30). The deflected beam then strikes one of the 1024 different holograms (optical interference patterns) in a flat array called a "hololens." The hololens splits the beam into two parts, one of which goes straight through while the other is diffracted to fall

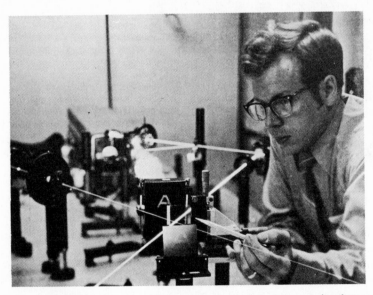

FIGURE 14.29. Writing and reading a hologram of the letter *A* simultaneously. The rear *A* is the original and the weaker, forward image is just being formed. (Courtesy of RCA.)

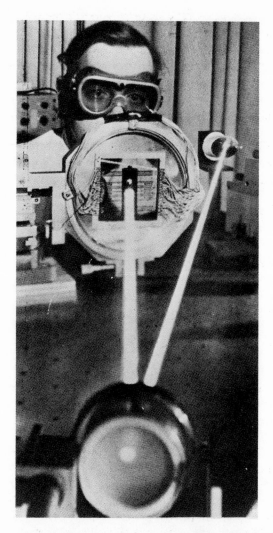

FIGURE 14.30. A laser beam is used to write, store, read, and erase information in an experimental, holographic, optical computer memory (courtesy of RCA).

on a flat plane 3-in. square that is composed of 1024 liquid crystal cells—tiny areas that can be made by electronic means to be scattering or transparent. These cells introduce digital information into the laser beam in the form of tiny areas that are dark (where the liquid-crystal cell is scattering) or light (where it is transparent). These correspond to the "ones" and "zeroes" of binary code and constitute the data to be written into the memory.

This coded, or modulated, "object" laser beam as well as the straight-through, or "reference," beam recombine at the selected location on the storage medium where they interfere and produce a stationary light-intensity pattern. The storage medium itself consists of a thermoplastic–photocon-ductor sandwich which has been charged by a corona to a few hundred volts. The photoconductor alters the potential across the thermoplastic in accord-ance with the light intensity, and the resulting differential electrostatic forces cause the thermoplastic to deform when heated. The deformations are a replica of the original light pattern, and upon cooling store the pattern as a phase-relief hologram.

Readout is achieved by sending the laser beam through the system so that it strikes the hologram selected for readout but not the liquid crystal array. The beam reconstructs the image of light and dark areas stored in the holo-gram and projects it onto an array of light-sensitive elements that produce the electronic equivalent of the data. Erasure is accomplished by heating the thermoplastic to a temperature somewhat higher than that used for writing. Once erased, the thermoplastic may be used over again. The experimental unit comprises $32 \times 32 = 1024$ pages, each with $32 \times 32 = 1024$ bits, for a total capacity of more than a million bits.

14.26. One-Dimensional Hologram Memories

Recently, a hologram–tape combination memory was described along the lines of the one-dimensional hologram zone plates discussed at length earlier.[70] Through a bit of manipulation of light paths, the approach eases the storing and reproduction of optical signals on tape or disks. Using holo-grams, the method offers high bit densities while eliminating the mechanical-tolerance problems associated with other optical recording and readout methods. The technique, developed at the Munich Laboratories of Siemens AG, uses one-dimensional holograms that are projected by laser beams onto an erasable recording medium—a thermoplastic, for example. For reading out the information contained in the holograms, the tape or disk is *moved past* a constant reference beam.

In experience performed so far, the Siemens approach has yielded a linear storage density of 1000 bits/mm along a 3-μ wide track, with the expectation being that it will be possible to produce 1.5- to 2-μ tracks, increasing thereby the packing density to 10^5 bits/mm^2.

H. Kiemle, of Siemens, has discussed the recording, on a continuously moving medium, of narrow, one-dimensional hologram tracks, analogous to the recording of moving magnetic tape.[71,72]

14.27. References

1. W. E. Kock, Aehnlichkeit zwischen Vokal-formanten und Formanten von Musik-Instrumenten, *ETZ* **5**, 166 (1953).
2. W. E. Kock, *Seeing Sound*, Wiley, New York, 1971, p. 58.
3. R. K. Potter, G. S. Kopp, and H. C. Green, *Visible Speech*, Van Nostrand, Princeton, N.J., 1947.
4. G. Runge, *Z. Math. Phys.* **XVII**, 443 (1903).
5. W. E. Kock, University of Cincinnati, E.E. Thesis, 1932.
6. J. Lipka, *Graphical and Mechanical Computation*, Wiley, New York, 1921, p. 184.
7. J. W. Cooley and J. W. Tukey, An algorithm for the machine calculation of complex Fourier series, *Math. Comp.* **19**, 297–310 (1965).
8. G. D. Bergland, A guided tour of the fast Fourier transform, *IEEE Spectrum* **6**, July, 41–52 (1969).
9. H. Stenzel, *Leitfaden zur Berechnung von Schallvorgaengen*, Springer, Berlin, 1939, p. 59.
10. K. Osterhammel, Optische Untersuchung des Shallfeldes Kolbenfoermig Quarze, *Akust. Z.* **6**, 82 (1941).
11. W. E. Kock, Unpublished Bell Telephone Laboratories memorandum, March 1948.
12. W. E. Kock and J. L. Stone, Space frequency equivalence, *Proc. IRE* **46**, 499–500 (1958).
13. W. E. Kock, Related experiments with sound waves and electromagnetic waves, *Akoustische Beihefte* **1**, 227–238 (1959) [also appeared in *Proc. IRE* **47**, 1192–1201 (1959)].
14. D. G. Tucker, Space-frequency equivalence in directional arrays, *Proc. IEE* **109C**, 191–197 (1962).
15. M. I. Skolnik, Application of space frequency equivalence to radar, paper presented at the IRE convention, 1962.
16. L. B. Lesem, P. M. Hirsch, and J. A. Jordan, Jr., The promise of the kinoform, *Opt. Spectra* **4**, December (1970), 18–21.
17. Lord Rayleigh, On the theory of optical images with special reference to the microscope, *Phil. Mag.* **42** (1896), 167–195.
18. R. W. Wood, *Physical Optics*, 3rd ed., Macmillan, New York, 1934, pp. 277–279.
19. K. Preston, Jr., *Electronics* **38**, 62 (1965).
20. S. P. McGrew, A proposed active optical computer, paper presented at the NATO Advanced Study Institute, June, 1973, Capri, Italy (Director Professor E. R. Cianiello).
21. E. Abbe, Beiträge zur Theorie des Mikroskops und der mikroskopischen Wahrnehmung, *Archiv Mikrosk. Anat.* **9**, 413–468 (1873).
22. A. Marechal and P. Croce, Un filtre de frequences spatiales pour l'amelioration du contraste des images optiques, *C. R. Acad. Sci.* **237**, 607–609 (1953).
23. L. J. Cutrona, E. N. Leith, C. J. Palermo, and L. J. Porcello, Optical data processing and filtering, *IRE Trans. Inf. Theory*, **IT6** (3), 386–400 (1960).
24. E. L. O'Neill, *Introduction to Statistical Optics*, Addison-Wesley, Reading, Mass., 1963.
25. M. B. Dobrin, Optical processing in the earth sciences, *IEEE Spectrum* **5**, Sept., 59–66 (1968).
26. M. B. Dobrin, A. L. Ingalls, and J. A. Long, Velocity and frequency filtering of seismic data using laser light, *Geophysics* **30** (6), 1144–1178 (1965).

27. G. W. Stroke, Etude theorique et experimentale de deux aspects de la diffraction par les reseaux optiques, *Rev. Opt.* **39**, 291–396 (1960).
28. G. W. Stroke, *An Introduction to Coherent Optics and Holography*, Academic Press, New York, 1969, pp. 3–4.
29. A. Vander Lugt, Signal detection by complex spatial filtering, *IEEE Trans. Inf. Theory*, **10**, 145–153 (1964).
30. J. Tsujiuchi, in: *Progress in Optics*, North Holland, Amsterdam, 1963, Vol. 2, p. 133.
31. E. N. Leith, A. Kozma, and J. Upatnieks, in: *Optical and Electrooptical Information Processing*, M.I.T. Press, Cambridge, Mass., 1965.
32. G. W. Stroke and R. G. Zech, A posteriori image correcting "Deconvolution" by holographic Fourier-transform division, *Phys. Lett. A* **25**, 89–90 (1967).
33. R. H. Dicke, *Astrophys. J.* **153**, 101–106 (1968).
34. G. W. Stroke, *Phys. Lett. A* **27**, 252–253 (1968).
35. G. W. Stroke, *Phys. Lett. A* **27**, 405–406 (1968).
36. G. W. Stroke, F. Furrer, and D. R. Lamberty, *Opt. Commun.* **July/Aug.**, 141–145 (1969).
37. S. G. McCarthy and I. Roth, *Sperry Eng. Rev.* **1**, 41–45 (1966).
38. E. N. Leith, Optical processing techniques for simultaneous pulse compression and beam sharpening, *IEEE Trans. Aerosp. Electron. Syst.* **4**, 879–885 (1968).
39. G. W. Stroke, Optical computing, *IEEE Spectrum* **9**, Dec., 24–41 (1972).
40. L. J. Cutrona, E. N. Leith, L. J. Porcello and W. E. Vivian, On the application of coherent optical processing techniques to synthetic-aperture radar, *Proc. IEE* **54** (8), 1026–1032 (1966).
41. W. E. Kock, Holographic computing in radar and ultrasonics, *IEEE Comput. Soc. Dig.* **Apr.**, (1972).
42. W. E. Kock, Parallel processing in synthetic aperture systems, in: *New Concepts and Technologies in Parallel Information Processing*, NATO Advanced Study Institute, Nordhoff International Publ., Groningen, The Netherlands, 1974 (Prof. E. R. Cianiello, ed.).
43. W. E. Kock, Optical computing in synthetic aperture radar, *Proc. Soc. Inf. Disp.*, **15/3** (3rd quarter), 112–118 (1974).
44. W. E. Kock, A real time parallel optical processing technique, *IEEE Trans. Comput.* (Special Issue on Optical Computing) **C-24**, Apr., 407–411 (1975).
45. K. Preston, Jr., Digital holographic logic, *Digest Opt. Comput. Symp. Darien, Conn.*, Apr. 1972.
46. J. Feinleib, Optical processing in real time, *Laser Focus* **9**, Sept., 42–44 (1973).
47. G. Marie, *Philips Res. Rep.* **22**, 119 (1967).
48. D. Casasent, Applications of a real time hybrid computing system, in: *IEEE Digest of Papers, 1974 International Optical Computing Conf., Zurich, Apr. 9–11, 1974*, pp. 18–22.
49. B. R. Brown and A. W. Lohmann, Complex spatial filtering with binary masks, *Appl. Opt.* **5** (6), 967–969 (1966).
50. B. R. Brown and A. W. Lohmann, Computer generated binary holograms, *IBM J. Res. Dev.* **13**, 160–168 (1969).
51. O. N. Tufte and D. Chen, Optical techniques for data storage, *IEEE Spectrum* **10**, Feb., 25–32 (1973).
52. L. K. Anderson, Holographic optical memory for bulk data storage, *Bell Lab. Rec.* **46**, 318 (1968).
53. R. J. Collier, C. B. Burckhart and L. H. Lin, *Optical Holography*, Academic Press, New York, pp. 476–488.

54. A. L. Mikaeliane, V. I. Bobrinev, S. M. Naumov, and L. Z. Sokolova, Principles of holographic memory devices, paper presented at the IEEE Conference on Laser Applications, May 26–28, 1969.
55. D. H. McMahon, Holographic Ultrafiche, *Appl. Opt.* **11**, 798–806 (1972).
56. *Electron. News* **16**, June 14, 38 (1971).
57. *Lasersphere* **1**, Nov. 15, 1 (1971).
58. *IEEE Spectrum* **10**, Nov., 34 (1973).
59. *Ind. Res.* **16** (6), 58 (1974).
60. E. N. Leith, A. Kozma, J. Upatnieks, J. Marks, and N. Massey, Holographic data storage in three-dimensional media, *Appl. Opt.* **5**, 1303–1311 (1966).
61. M. R. Tubbs, Holographic storage, *Opt. Spectra* **7**, Apr., 5–10 (1973).
62. J. J. Amodei and D. R. Bosomworth, Hologram storage and retrieval in photochromic strontium titanate crystals, *Appl. Opt.* **8**, 2473 (1969).
63. J. C. Urbach and R. W. Meier, Thermo-plastic Xerographic holography, *Appl. Opt.* **5**, 666 (1966).
64. J. T. LaMacchia, Holographic storage in ferro electrics, paper presented at the Joint IEEE CVA Symposium on Applications of Ferroelectrics, Oct. 1968.
65. *Ind. Res.* **15** (11), 21 (1973).
66. *Laser Focus* **10**, Feb., 30 (1974).
67. The Fifth Holography School, Feb. 1973, held at Akademgorodok, Novosibirsk, Siberia.
68. *Lasersphere* **1**, Nov. 15, 1 (1971).
69. *Opt. Spectra* **7**, Apr., 24–25 (1973).
70. *Electronics* **47**, Sept. 5, 53 (1974).
71. H. Ruell and H. Kiemle, *Opt. Commun.* **7**, 158 (1973).
72. H. Kiemle, Considerations of holographic memories in the gigabyte region, *Appl. Opt.* **13**, 803–807 (1974).

15

NUCLEAR FUSION

When C. L. Townes, A. M. Prokhorov, and N. Basov conducted their first experiments on the physical process of stimulated emission of radiation, little did they suspect that their brainchild-to-be, the laser, might someday become the solution for the world's energy problems. Even when, two decades later, in 1964, the importance of their experiments was given world recognition through the announcement of their Nobel Prize award, the possibility of the laser acting as the key component in a nuclear power generating plant was just an idea being speculated upon, But when one of the three, Basov (Figure 15.1), reported on some experiments in 1968 in which he used a powerful laser as a tool in attempts to achieve a controlled thermonuclear fusion reaction, scientists began to take notice. When, in 1970, the French Atomic Energy Commission made the announcement that, at their laboratory at Limeil outside of Paris, repeated laser-induced fusion reactions had indeed occurred (Figure 15.2), the bandwagon started. Today, in many laboratories, here and abroad, large laser-induced-fusion programs are underway, including nearly a dozen in the U.S. alone.

On April 19, 1972, the U.S. Atomic Energy Commission declassified some key laser-induced fusion concepts, thereby disclosing to the public how close researchers were to achieving a reaction referred to as an exoergic one (one for which the nuclear power generated is larger than the power used to drive the laser). Accordingly, the concept of a laser-driven, nuclear-fusion power generating station can be looked upon as a key contender for the world's ideal energy source. Laser-induced fusion reactions use forms of hydrogen as their fuel, and hydrogen is an extremely plentiful component in the oceans of the world.

FIGURE 15.1. Nobel Laureate Nikolai Basov (left) discusses his latest, 1973, laser-fusion experiments with the author in Basov's office at the Lebedev Institute in Moscow.

15.1. Power Needs

In recent decades the world's energy needs have grown enormously, so that the eventual depletion of the world's coal, gas, and oil supplies has rapidly come dangerously near. Accordingly, the search for new ways for generating power by atomic means is an area of technology of prime importance to all nations. Since the discovery, some three decades ago, of a way for generating electrical power through nuclear processes, several hundred electric power generating stations utilizing atomic energy have been built throughout the world, with many more presently under construction.

In contrast to oil- or coal-fired electric power stations, nuclear stations use the metal uranium as their "fuel," and because the energy developed by them is derived from an *atomic* reaction, the needed amount of this fuel is quite tiny compared to the equivalent amount of coal or oil. Fuel transportation costs are therefore minimal. However, the world's supply of uranium is also, of course, limited, and even though only small amounts of it are used, nuclear power stations may also someday face the situation where the world's supply of uranium is depleted. A recently developed form of reactor, called the *breeder* reactor, because it breeds (generates) atomic fuel, will alleviate to

FIGURE 15.2. A section of a 1-mm-diameter wire of frozen-solid deuterium becomes, when a focused laser beam of 4×10^9 W strikes it, a tiny hydrogen bomb, thereby holding promise for an energy source using the plentiful water of the oceans as a "fuel" (courtesy of Hadron, Inc.).

some extent this problem of fuel depletion. The breeder reactor can thus become the next step in reactor design, but many scientists and engineers are already working diligently on adapting the *fusion* process to reactor technology since it can utilize the very plentiful sea water for its source of fuel.

15.2. Atomic Energy Development

The story starts in 1903 with Einstein's analysis which showed that energy and mass are equivalent, and that they are related in a very interesting way. Einstein's equation

$$E = mc^2 \tag{17.1}$$

states that energy E and mass m are proportional with the constant of proportionality being the square of a very large number, the velocity of light, c. It was thus recognized very clearly that if even very tiny amounts of mass could be converted into energy, the resulting energy would be exceedingly large.

As physics laboratories throughout the world continued to probe the atom through bombardment with faster and faster electrons and ions, the concept of successfully "splitting the atom" soon became the popular goal. This goal was recognized as being extremely important because of the manner in which the masses of the elements in the periodic table vary. Thus, for example, the nucleus of the helium atom (called an alpha particle) consists of two protons and two neutrons, whereas a deuteron (the nucleus of deuterium, a form of heavy hydrogen), consists of one proton and one neutron. But the *mass* of two deuterons is *more* than the mass of an alpha particle. Hence, if the two deuterons could be made to fuse, energy would be released, in an amount equal to the mass discrepancy. Such a process is called thermonuclear *fusion*; it is the process which is the basis for the huge-energy-releasing hydrogen bomb.

In 1938, the German scientists, O. Hahn and F. Strassmann, reported the splitting of a uranium atom (nucleus) into two fragments through neutron bombardment. In that instance, the combined mass of the fragments was *less* than that of the uranium nucleus. During the splitting process, this difference in mass appears as energy [about 200 MeV (million electron volts) energy for every uranium nucleus fragmented]. For each such *fission* process, several neutrons are released. Since the original splitting is caused by neutron bombardment, it was soon recognized that the new, released neutrons could cause still other nuclei to undergo fission, thus generating a *chain reaction*.

15.3. Nuclear Weapons

On December 2, 1939, Einstein wrote to President Roosevelt pointing out the possibility of the fission process being used to build weapons perhaps a million times more powerful than those then available and calling attention to a group of researchers working under Enrico Fermi who were devoting their efforts to the nuclear fission process. Support was given, and on December 2, 1942 Fermi's first experimental nuclear reactor came into being at the University of Chicago. In that reactor, as in all of today's nuclear reactors, whether they provide electric power to the public or power for propelling ships and submarines, the chain-reaction *fission* process just described is utilized. Whereas such a chain reaction, if *uncontrolled*, leads to a very rapid, explosive, result, this reaction can easily be *moderated*, permitting only a

FIGURE 15.3. An experimental nuclear explosion involving the original form of atomic bomb, the *fission* bomb.

desired number of fission processes to occur per unit of time, with a resulting constant power output. For nuclear weapons, on the other hand, an extremely rapid chain reaction and energy buildup is desirable, so as to effect maximum utilization of the material undergoing fission before the explosive force causes it to separate to distances for which continued neutron splitting is impossible. The uranium isotope U-235 (which is present in natural uranium in quantities of only 0.7%), and the transuranium element plutonium-239 (not available in nature but formed in fission reactors), are two materials possessing this needed property. On July 16, 1945 the first nuclear weapon was exploded, near Alamagordo, New Mexico, having an explosive power equivalent to 20,000 tons of TNT (Figure 15.3).

In 1939 Hans Bethe hypothesized that the *fusion* of hydrogen into helium might be one of the important sources of *stellar* energy. In this process, the light nuclei at the low end of the periodic table are fused together to produce heavier and more tightly bound nuclei. As we noted, the "lost mass" at this low end of the table results in a release of energy equivalent to that mass. Under the leadership of Edward Teller, U.S. scientists developed this concept into a weapon, and on November 1, 1952, the first "hydrogen" weapon (now called thermonuclear) was demonstrated near Eniwetok in the Marshall Islands of the Pacific (Figure 15.4).

15.4. Fission Reactors

Meanwhile, the great value as a power source, of Fermi's early contribution to the nuclear weapon program, the *controlled* fission process of a nuclear

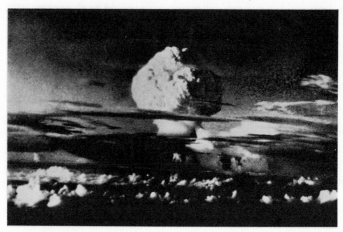

FIGURE 15.4. The experimental explosion of the first U.S. hydrogen (thermo-nuclear) bomb, occurring on November 1, 1952 at Eniwetok.

reactor, was becoming recognized, and the U.S. Atomic Energy Commission and U.S. industry began to consider ways for using reactors to generate electric power. In December 1951 the first lighting of an electric light bulb with power generated by a nuclear reactor occurred at the National Reactor Testing Station in Idaho (Figure 15.5). The Navy also began developing reactors that could be used to power ocean-going ships. The first nuclear-powered vessel was the submarine Nautilus. Soon, thereafter, on July 18,

FIGURE 15.5. The first lighting of electric light bulbs using power generated by a nuclear reactor occurred in December 1951.

1955, surplus steam produced during tests of the prototype reactor for another nuclear submarine, the Seawolf, was fed to a turbine generator, and the electricity thus generated was then distributed on the power lines of the Niagara-Mohawk Power Corporation, providing for the first time, electricity generated by nuclear power for *commercial* use. Commercial power plants utilizing nuclear energy began to materialize and then to proliferate, and during the period 1966–1968 plans for 50 nuclear power stations were formulated, with a total power output of 500 million watts, equivalent to one third of the total U.S. power-generating capability which existed in 1966.

15.5. The Breeder Reactor

Progress did not stop with the development of "standard" nuclear reactors. As we noted, uranium, like coal and oil, will face eventual world depletion. Accordingly, the concept of the *breeder* reactor, in which fuel is generated (bred), was looked upon as an important goal. Fermi, in 1945, stated "The first nation to develop a successful breeder reactor will gain a tremendous economic advantage." The U.S. Atomic Energy Commission assigned this task to the Argonne National Laboratory, located on the outskirts of Chicago. Today, their very successful EBR-II liquid-metal breeder reactor (located near Idaho Falls, Idaho) has shown that such a reactor should

FIGURE 15.6. The EBR-II, the first U.S. liquid-metal *breeder* reactor, developed by the Argonne National Laboratory and located near Idaho Falls, Idaho, *breeds* nuclear fuel as it generates electric power.

be commercially feasible (Figure 15.6). Accordingly, in August 1972, a memorandum of agreement was signed for the construction of the first U.S. commercial liquid-metal breeder reactor, with its construction intended to demonstrate the importance of breeder reactors as potential sources of electrical power. The power station, presently under construction, is financed by the Tennessee Valley Authority, the Atomic Energy Commission, Commonwealth Edison, and a consortium of private utilities.

15.6. Controlled Fusion

Now fusion is next. It has long been recognized that if the thermonuclear or fusion process could be *controlled* (instead of uncontrolled, as in the hydrogen bomb) many benefits would accrue. In 1956, Amasa B. Bishop, then Chief of the Controlled Thermonuclear Branch, Research Division, U.S. Atomic Energy Commission, pointed out three advantages of a reactor using the controlled thermonuclear reaction (CTR). First, it would contain within the reaction chamber only a minute amount of "fuel" at any one time, so that the possibility of an explosion or loss of control would be almost nil. Second, the fusion process, when the nuclear action is carried to proper completion, would avoid the problem faced by today's fission reactors, that of disposing of the relatively useless, spent, highly radioactive, fission products. Finally, one of the likely fuel-element possibilities, deuterium, is a naturally occurring isotope of hydrogen with an abundance in sea water of about one part in 6000, so that there is enough deuterium in ocean water to provide all of mankind's present power requirements for several hundreds of millions of years.

Controlled fusion was given much effort in the last two decades, and although progress of a sort was made, the picture still looked very distant, when, in 1970, the French Atomic Energy Commission announced positive evidence of having repeatedly generated a controlled thermonuclear reaction, using a powerful laser as the ignition device (Figure 15.7). The fusion process requires that the elements to be fused be brought to an exceedingly high temperature, and the achieving of such high temperatures was always recognized by those working in the controlled thermonuclear field as one of the most formidable difficulties facing them. However, soon after the laser was first demonstrated (in 1960), experimenters were finding that its very special coherent light could be focused into extremely tiny focal areas, thereby concentrating its energy to an unbelievable degree, and generating temperatures high enough to punch (burn) holes in razor blades (Figure 6.6).

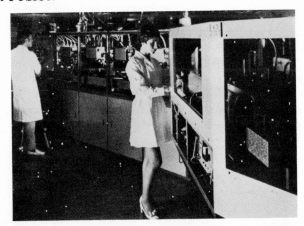

FIGURE 15.7. The Compagnie Generale Electrique ultrabrightness VD–450 laser which produced the reaction in Figure 15.2. In the background are the laser and four amplifier stages.

15.7. Amplifying the Laser Pulses

We have seen how the Q-spoiling procedure can increase both the number of stimulated light emissions and the suddenness with which they occur. Both of these effects result in an increase in the power in the pulse, since for the same light energy, the shorter time of generation means more energy per second, and hence more power.

After the Q-spoiling process has been carried to its limit, laser amplifiers can be used to increase the pulse power even further. As we noted, such amplifiers are very similar to the (pulse-generating) laser, having a flash of light acting as a pumping source and a rod of laser material; for them, however, the reflectors of the laser of Figure 6.3 are eliminated. In this arrangement, the light flash generated for the amplifier must be accurately timed so that a significant population inversion exists in its rod when the laser pulse arrives. The emissions from the amplifier rod (stimulated by the exactly correct wavelength light in the pulse) add coherently to the incoming pulse and a much more powerful pulse emerges from the amplifier rod. In addition, the leading edge of the pulse usually receives most of the additional energy, so that the amplified pulse is *shorter*. This process can be repeated by using a series, or chain, of amplifiers. Eventually, however, the power in the pulse becomes so great that the rod in the final amplifier is shattered by it.

At this point other procedures must be resorted to. One technique employed by French laser designers resulted from their observation that the rods in the final amplifier always failed at its outside surface, i.e., at its perimeter. This is because the pulse is amplified more along the outer area of the rod, this being due to its being closer to the pumping light from the external coiled flash lamp. So, to more effectively utilize the entire volume of the last amplifier rod, a spatial filter is employed at the output of the initial laser, which concentrates a certain portion of the energy into the central section of the first amplifier rod. When this spatially "shaped" pulse finally reaches the exit point of the last amplifier, the outer-edge portions of it, having been amplified more, again become equal in amplitude to the central portion, and a spatially uniform pulse is the result. We saw a typical amplifier chain in Figure 7.1.

Another procedure for increasing the pulse power involves the use of parallel amplifier chains. Early in the amplifying process the pulse is split in two along its central vertical plane, and each half is then reshaped and fed into its own amplifier chain. The last amplifier rod of each of the chains is again operated at its maximum value, just short of shattering, and when the output pulses are combined, the maximum power is thus doubled. Basov, at the Lebedev Institute, in Moscow has set up a laser system having many such parallel amplifier chains, permitting an increase in the maximum pulse power by a sizable factor. This paralleling of amplifiers also permits the fusion material to be illuminated simultaneously from many different directions.

15.8. Laser-Fusion Experiments

In May 1968, Basov reported on his experiments conducted at the Lebedev Institute in Moscow. He employed a neodymium-glass laser having 5 amplifier stages and used lithium deuteride as his target. His equipment generated a 30-J laser pulse 200-nsec duration. His results were somewhat uncertain; he noted, however, that in about every sixth experiment he detected a few neutrons. During the Summer of 1969, similar experiments, using a somewhat more powerful laser pulse, were conducted at the Sandia Laboratories of the U.S. Atomic Energy Commission, with slightly better results.

The announcement which attracted large world-wide attention came in 1970 from the French Atomic Energy Laboratories at Limeil (the Limeil Weapons Research Center). A 1967 joint effort between that organization and the French firm Compagnie Generale Electrique (C.G.E.) had led to the development of an especially pure form of neodymium glass for use in laser rods. This neodymium-doped glass could survive 40 J/cm^2 without breaking, more than double the resistance of similar glass then available in the U.S. Using this glass, C.G.E. was able to construct a laser amplifier chain which could generate 80 J in 2 nsec (as compared to the Basov 200-nsec-lower-energy laser pulse). When this laser was used at the Limeil laboratory to bombard a cylinder of frozen heavy hydrogen (deuterium) in a long set of tests, close to a thousand neutrons were generated with each pulse (Figure 15.2). One of the important fusion reactions involves only the quite plentiful form of hydrogen, deuterium, and is referred to as the D-D reaction. In this fusion of hydrogen to form helium, it was known that neutrons are generated having an energy equivalent to 3.3 MeV. At Limeil, the energy of the neutrons observed was determined through time-of-flight travel measurements over a known distance; these measurements confirmed their assumption that the D–D fusion reaction had indeed been made to occur.

The Limeil announcement made headlines in science journals the world over. Figure 15.2 appeared in color on the cover of the issue of *Laser Focus*, and since that time, many articles discussing laser fusion have appeared.[1-15] Nevertheless, many remained highly skeptical of the achieving, within any reasonable time, of the ultimate goal, the generation of an amount of nuclear power significantly larger than that used to power the laser.

15.9. Pulse Tailoring

In April 1972, the U.S. Atomic Energy Commission declassified certain key concepts regarding an implosion effect which can be shown analytically to occur when a fusion fuel pellet is irradiated by a properly "time-tailored" laser pulse. This declassification step enabled Dr. Keith Bruckner, of KMS Industries in Ann Arbor, Michigan, to give two days later, on April 21, 1972, in a previously scheduled lecture to a packed hall at the Chrysler Auditorium of the University of Michigan, details of *his* implosion procedure. Shortly thereafter, at the Quantum Electronics Conference held at Montreal, Canada, in May 1972, several postdeadline papers (which had previously been unsubmittable because of classiffication) were presented by Nuckolls and his associates at the U.S. Atomic Energy Commission Laboratory, the Lawrence Livermore Laboratory, in California.[1] In these papers, a second form of pulse tailoring for achieving a sizable implosion effect was discussed. Implosion increases the density of the pellet; this provides a significant advantage because the likelihood of the nuclear reaction varies as the *square* of the density of the pellet. Experiments at KMS in 1974 showed visible volume compressions of 88, with the *actual* compressions presumably much higher.[2] In 1975, Battelle Laboratories, Columbus, reported the successful operation of a Nd : glass laser-amplifier system capable of delivering 1400 J in a 3.5 nsec. pulse.

15.10. Uranium Enrichment By Lasers

As a final note on the use of lasers in the nuclear energy field, the possibilities of uranium enrichment should be mentioned. The biggest obstacle in preparing uranium for a nuclear fission reactor is the tedious, costly process of separating the active isotope U-235 from its more plentiful, inert sister U-238. Since the two isotopes have identical chemical properties, this enrichment process has traditionally relied on the tiny difference in *weight* between atoms of the two substances to separate them, a process called gaseous diffusion. By the 1980's present gaseous diffusion plants will simply not be able to keep up with the growing demand for reactor-quality uranium, and intensive research is going on to find other methods of separation. In experiments in laser separation conducted at the Lawrence Livermore Laboratory, the beam from a dye laser tuned to 5915.4 Å was directed through a stream of vaporized uranium heated to 2500°K. The laser photon energy matches a transition frequency for bringing U-235, but not U-238, from its ground state to an excited state. Ultraviolet light from a mercury arc lamp is then used to *ionize* the excited U-235 atoms, permitting them to be collected electrostatically. In the experiments, the 0.7% natural concentration of U-235 was increased to 60%; unfortunately, it was estimated that full-scale operation of photoionization plants for uranium enrichment would require the development of lasers three orders of magnitude more powerful than those presently available. More recently, however, it was announced by the Livermore Laboratory that even with present-day lasers, the enrichment rate had been "inceased 10-million-fold over the previously reported one."[18]

Other research teams are conducting similar experiments at the Tel Aviv University and at an Exxon–Avco cooperative laboratory, with this latter group reportedly being closest to a pilot-plant design.

15.11. References

1. J. Nuckolls, J. Emmett, and L. Wood, Laser-induced thermonuclear fusion, *Phys. Today* **26** (8), 46–53 (1973).
2. KMS claims thermonuclear neutrons from laser implosion, *Phys. Today* **27** (8), 17–19 (1974).
3. New laser may lead to fusion, *Ind. Res.* **15**, 21–22 (1973).
4. Laser-fusion claim: An evaluation, *Sci. News* **May 25**, 333 (1974).
5. P. E. Thomsen, Laser fusion: On the road to laser fusion power at Lawrence Livermore Laboratory, *Sci. News* **Aug. 17**, 106–111 (1974).
6. L. R. Solon, Lasers and fusion, *Ind. Res.* **13** (1971).
7. Nixon expected to accelerate laser-fusion effort by 33%, *Laser Focus* **10**, Feb., 10–13 (1974).

8. N. Basov and O. Krokhin, Laser light for fusion fire, *New Sci. and Sci. J.* **Sept. 30,** 733 (1971).
9. On the way to fusion power, *Electronics* **46**, Sept. 27, 38 (1973).
10. *Ind. Res.* **16** (8), 19–21 (1974).
11. *Phys. Today* **27** (9), 17–20 (1974).
12. Enrichment by laser, *Sci. Am.* **231** (4), 57–58 (1974).
13. L. Bush, Lasered fuel, *Ann Arbor News* **Sept. 22**, 33 (1974).
14. C. D. Cantrell, Uranium enrichment, *Science (Lett.)* **Sept. 27**, 1109 (1974).
15. Uranium enrichment by laser achieved, *Sci. News* **June 22**, 396–397 (1974).
16. Passing a laser fusion milestone, *Ind. Res.* **17**, 24 (1975).
17. Laser fusion: one milepost passed, *Science*, **Dec. 27** (1974).
18. *The Wall Street Journal*, **LV**, p. 7, May 30, 1975.

FURTHER READING SUGGESTIONS

G. W. Stroke, *An Introduction to Coherent Optics and Holography*, Academic Press, New York, 1969, 2nd ed.

R. J. Collier, C. B. Burckhardt, and L. H. Lin, *Optical Holography*, Academic Press, New York, 1971.

E. Camatini, ed., *Optical and Acoustical Holography*, Plenum Press, New York, 1972.

H. Kiemle and D. Ross, *Einfuehrung in die Technik der Holographie*, Akademische Verlagsgesellschaft, Frankfurt a. M., 1969.

B. P. Hildebrand and B. B. Brendon, *An Introduction to Acoustical Holography*, Plenum Press, New York, 1972.

E. S. Barrekette, W. E. Kock, T. Ose, J. Tsujiuchi, and G. W. Stroke, eds., *Applications of Holography*, Plenum Press, New York, 1971.

W. E. Kock, *Lasers and Holography*, Doubleday, Garden City, N.Y., 1969.

W. E. Kock, *Seeing Sound*, Wiley, New York, 1971.

L. Goldman, *Applications of the Laser*, CRC Press, Cleveland, Ohio, 1973.

F. A. Jenkins and H. E. White, *Fundamentals of Optics*, McGraw-Hill, New York, 1957.

A. F. Metherell, H. M. A. El-Sum, and L. Larmore, eds., *Acoustical Holography*, Plenum Press, New York, 1969, Vol. I.

A. F. Metherell and L. Larmore, eds., *Acoustical Holography*, Plenum Press, New York, 1970, Vol. II.

A. F. Metherell, ed., *Acoustical Holography*, Plenum Press, New York, 1971, Vol. III.

Glen Wade, ed., *Acoustical Holography*, Plenum Press, New York, 1973, Vol. IV.

P. S. Green, ed., *Acoustical Holography*, Plenum Press, New York, 1974, Vol. V.

W. T. Cathey, *Optical Information Processing and Holography*, Wiley, New York, 1974.

D. D. Dudley, *Holography: A Survey*, NASA Technology Utilization Office, U.S. Government Printing Office, Washington, D.C., 1974.

M. Ross, *Laser Applications*, Academic Press, New York, Vol. I, 1971; Vol. II, 1974.

H. A. Elion, *Laser Systems & Applications*, Pergamon Press, New York, 1967.

R. J. Bell, *Introductory Fourier Transform Spectroscopy*, Academic Press, New York, 1972.

C. S. Williams and O. A. Becklund, *Optics, A Short Course for Engineers and Scientists*, Wiley, New York, 1972.

B. A. Lengyel, *Lasers*, Wiley, New York, 1971. 2nd ed.

M. S. Field, A. Javan, and N. A. Kurnit, eds., *Fundamental and Applied Laser Physics*, Wiley, New York, 1971.

G. W. Stroke, W. E. Kock, Y. Kikuchi, and J. Tsujiuchi, eds., *Ultrasonic Imaging and Holography in Medical, Sonar, and Optical Applications*, Plenum Press, New York, 1974.

W. E. Kock, *Radar, Sonar, and Holography*, Academic Press, New York, 1973.

W. E. Kock, Acoustic holography, in: *Physical Acoustics*, Academic Press, New York, 1973, Vol. 10.

W. E. Kock, Microwave holography, in: *Holographic Non-Destructive Testing*, Academic Press, New York, 1974.

AUTHOR INDEX

391

SUBJECT INDEX

Aberration in lenses, 187
Acceleration measurement, 154
Acoustic holograms, 70
Acoustic intensity, 324
Acoustic lens, 5
Advanced Research Projects Agency, 366
Aero Service Corporation, 205
Aerospace Corporation, 134, 137
AGFA, 98
Air-speed indicator, 153
Alamagordo, 379
Aldrin, Buz, 151
ALGOL, 345
Alignment applications, 145, 146
Amazon River basin, 204, 205
Anderson Hospital, 323, 324
ANIK, 298
Anti-Stokes line, 165
Apollo 11, 131
Apollo 15, 151
Argonne National Laboratory, 381
Armstrong, Neil, 151, 296
Aurora borealis, 108
AVCO—Everett, 272

Babinet's principle, 212, 213, 215
Bahamas, 238
Bandwidth, 25
Bangalore, 67
Battelle Laboratories, 385
Bell Telephone Laboratories, viii, 128, 130,
   136, 139, 140, 141, 158, 183, 189,
   251, 255, 274, 275, 291, 292, 299,
   305, 334, 365

Bendix Corporation, viii, 85, 111, 228, 234,
   239, 266, 319, 351
Bermuda, 238, 239, 240
Bistatic systems, 215, 237
Black Sea, 263
Boeing, 147
Boulder, Colorado, 149
Breeder reactor, 376, 381
Brewster's angle, 115, 128
Bright spot in the shadow of a disk, 9, 10
Brillouin spectroscopy, 164, 165
British National Physical Laboratory, 176
Brownian movement, 175
B-scan, 316, 317, 318

Canadian Defense Research Establishment,
   130
Canadian Research and Development, 166
Car speed, 152
Carbon dioxide laser, 129
Carrier injection, 136
Cartier's, 78, 79, 286
Chain reaction, 378
Chalcogenides, 367, 369
Chemical laser, 133
Chirp, 255, 257, 265
Chromatographic analysis, 288
Circular waves, 7, 8, 9, 29
Clear air turbulence, 156
Coaxial cable, 291
Coherence, 2
Coherence length, 118
Coherent radiation, 140
Columbia University, 120